Foreign Ministries and the Information Revolution

I0885544

Diplomatic Studies

Series Editor

Jan Melissen

Netherlands Institute of International Relations 'Clingendael'

VOLUME 2

Foreign Ministries and the Information Revolution

Going Virtual?

By

Jozef Bátora

MARTINUS
NIJHOFF
PUBLISHERS

LEIDEN • BOSTON
2008

This book is printed on acid-free paper.

Library of Congress Cataloging-in-Publication Data

Bátora, Jozef
Foreign ministries and the information revolution : going virtual? / by Jozef Batora.
p. cm.
Includes bibliographical references and index.
ISBN 978-90-04-16900-5 (hardback : alk. paper) 1. Foreign offices—Information technology—Case studies. 2. Diplomacy—Information technology—Case studies. 3. Internet in public administration—Case studies. 4. Diplomacy—History. I. Title.
JZ1410.B38 2008
353.1'32380285—dc22

2008016004

ISSN 1872-8863
ISBN 978 90 04 16900 5

Copyright 2008 by Koninklijke Brill NV, Leiden, The Netherlands.
Koninklijke Brill NV incorporates the imprints Brill, Hotei Publishing, IDC Publishers, Martinus Nijhoff Publishers and VSP.

All rights reserved. No part of this publication may be reproduced, translated, stored in a retrieval system, or transmitted in any form or by any means, electronic, mechanical, photocopying, recording or otherwise, without prior written permission from the publisher.

Authorization to photocopy items for internal or personal use is granted by Koninklijke Brill NV provided that the appropriate fees are paid directly to The Copyright Clearance Center, 222 Rosewood Drive, Suite 910, Danvers, MA 01923, USA.
Fees are subject to change.

PRINTED IN THE NETHERLANDS

To Zuzana and Adam

CONTENTS

ACKNOWLEDGEMENTS

Thanks are due to a number of people and institutions for making this book possible. First and foremost, I would like to thank John Erik Fossum and Harald Baldersheim. During the four years when an earlier version of this text was in the process of becoming a PhD dissertation at the Department of Political Science at the University of Oslo, John and Harald were perceptive readers, tough critics and inspiring intellectual guides. Our discussions helped me structure my thoughts not only about this book but also about life as an academic. The Department provided me with a stimulating research environment and a friendly collegial atmosphere where one felt fine irrespective whether one sat together with a brownbag-lunch in the 9th floor at Blindern or with a glass of tasty Italian wine during an 'institute-seminar' in the streets of Trastevere. I also benefited greatly from my stay as a visiting scholar at SCANCOR, Stanford University in the academic year 2003/2004, where I had the opportunity to discuss some of the ideas presented here with James G. March, Walter W. Powell, W. Richard Scott, Tom Christensen, Ivar Bleiklie, Amir Sasson and a number of other Scandinavian and "Southern-Scandinavian" organization theorists. I would also like to thank Johan P. Olsen for challenging and enlightening conversations on the nature of institutions, and to Elin Allern, Jeff Checkel, Morten Egeberg, Per Laegreid, Paul Roness, Ulf Sverdrup and Jarle Trondal for well targeted thoughts and comments on various papers, chapters and concepts I worked with while writing this. My brief but intensive stunt as senior researcher at ARENA at the University of Oslo in 2006 had also given the opportunity for sharpening some of the arguments presented here and in other parts of my work. Many thanks also to Janne Haaland Matláry and Audun Offerdal for their encouragement and support.

Among the students of diplomacy, I am particularly grateful to Brian Hocking for taking time to read through an earlier version of this text and making a host of useful comments. I have learned a lot from our numerous debates in recent years. I would also like to thank Iver B. Neumann for introducing me to the world of theoretical study of diplomacy and for interesting ideational exchanges. Many thanks also to Jan Melissen for his comments on earlier think-pieces addressing some of the issues discussed in this book.

A number of contacts in each of the three foreign ministries investigated here were helpful in providing guidance during my fieldwork. In particular I would like to thank Tore Nedrebø of the Norwegian Ministry of Foreign Affairs, Peter Holásek of the Slovak Ministry of Foreign Affairs, and Evan H. Potter of the Canadian Department of Foreign Affairs and International Trade.

I am indebted to the Institute for European Integration Research at the Austrian Academy of Sciences in Vienna, where this book was finalized. The colleagues there—in particular Isabella Eiselt, Monika Mokre, Johannes Pollak and Peter Slominski—provided a fine intellectual environment be that in the office or during the occasional lunch at the 'Naschmarkt'. I would also like to thank Silvia Miháliková of the Institute of European Studies and International Relations at Comenius University in Bratislava for her support.

Most of all, I am grateful to my wife, Zuzana Vahančíková-Bátorová, for seeing things I don't see and walking this world with me.

CHAPTER ONE

INTRODUCTION

Diplomacy in its modern form has evolved as a key institution of the Westphalian state order; i.e. diplomacy can be conceived of as a system of rules, norms, routines and procedures regularizing interaction and official communication between states. Diplomats from different countries operate based on shared codes of conduct and practices which have evolved over centuries of interstate interactions and indicate what are legitimate diplomatic actors, situations and actions. This shared institutional norm-set has been providing meaning, predictability and stability to interstate interactions and has been maintained by diplomats' prerogative as exclusive guardians of the system of state-to-state communication. Organizationally, the system is embedded in the global group of foreign ministries which, as Hocking (2007:14) suggests, are ideal-typically and traditionally perceived to work as filters through which messages pass between the domestic and the foreign environment; as repositories of skills in terms of policy advice on international issues; and as the respective loci of national institutional memory in the conduct of foreign affairs.[1] Today, due to the set of developments commonly referred to as the *information revolution*, this institutionalized system of inter-state communication is under strain. Easy availability of advanced information technology (IT) equipment and low costs of communication and information transfer have been increasingly empowering non-governmental organizations, local governments, private enterprises, universities and individuals in recent decades, so that the Canadian scholar James Eayrs could observe already in the early 1970s that "[t]he requirements for setting up your own department of external affairs in your basement are remarkably modest. You need only be reasonably literate, fairly persistent, moderately affluent" (Eayrs 1971:78). Foreign

[1] The role of a foreign ministry as an information bank on the state's foreign affairs is of particular importance, because "without the capacity to relate myriad past commitments and treaties to the present, and to each other, decision-makers would be left floundering in chaos, given the complexity of the contemporary international system (Hill 2003:77, cf. Hocking 2007:7).

ministries structured around and aimed at administering state-to-state interactions according to long-established procedures and routines are increasingly unable to handle the sheer volume, speed and proliferating formats of foreign-policy-relevant information coming from and exchanged between non-state actors irrespective of state borders. Given these developments, claims of state prerogatives in foreign affairs are perceived as difficult to sustain and diplomacy, the centuries-old set of practices and habits sustaining the stability of interactions between states, is said to be challenged.[2] The current study is an attempt to assess the nature of that challenge.

1.1 *IT-revolution and Diplomacy*

A revolution can be defined as "any major social and political transformation, sufficient to replace old institutions and social relations, and to initiate new relations of power and authority".[3] The set of developments commonly referred to as the information revolution based on cheap inputs of information, plummeting costs of information transmission and radically increasing computing capacities is part of a wider change dynamics in the international environment following the end of the Cold War which challenges the role of the state as the primary actor in the international system. When analyzing the impacts of the set of developments often referred to as globalization, some authors go as far as to announce the end of the nation state (e.g. Wriston 1992, Guehenno 1995). More moderate voices hold that states continue to exist, but no longer as centres of power, but rather as nodes of a broader multidimensional network of power where the other nodes are non-state actors such as multinational corporations, NGOs, supranational governance

---

[2] One of the most influential reports dealing with the topic, *Reinventing Diplomacy in the Information Age* (Washington, DC: Center for Strategic and International Studies, November 1998) argues that due to technological and other developments related to globalization, the "Westphalian world in which modern diplomacy was born is no longer recognizable." Other reports arguing for the need to 'reinvent' diplomacy along similar lines include *Equipped for the Future: Managing U.S. Foreign Affairs in the 21st Century*, Washington, D.C.: Henry L. Stimson Center, October 1998; *America's Overseas Presence in the 21st Century* (http://www.911investigations.net/IMG/pdf/doc-247.pdf)

[3] "Revolution" *The Oxford Dictionary of Philosophy*. Simon Blackburn. Oxford University Press, 1996. *Oxford Reference Online*. Oxford University Press. <http://www.oxfordreference.com/views/ENTRY.html?subview=Main&entry=t98.e2073>

structures, international organizations, regional governments, terrorist networks, criminal syndicates etc. (Castells 1997, Rosenau 1997). As Nye (2002:64) argues, the key issue for research is not the continued existence of the sovereign state, but how its centrality and functions are being altered. Held et al. (1999:85–86) see this situation as partly a realization of what Hedley Bull had called *a neo-medieval order*, when they observe that although political community and political space continue to be shaped by the reach of territorial state sovereignty, it is not exclusively so.[4] In the interpretation of Ulrich Beck (1993:214–217), there is a metamorphosis of the state from *Handlungsstaat* into *Verhandlungsstaat* where the state previously acting on a sovereign basis is now forced into continuous negotiation and legitimization of its position in relation to non-state actors. The challenge with such emerging patterns of diplomatic negotiation and representation is related to the fact that many of the new diplomatic issue areas and actors have not developed any form of effective governance and are hence unable to represent themselves (Langhorne and Wallace 1999:17). In such a situation, foreign ministries often find themselves in a situation where they negotiate on behalf of or in relation to fragmented and multidimensional groups of non-state actors, which presents numerous procedural challenges (Price 1998, Neumann 1999, Axworthy 2003) and questions about the appropriate role of the foreign ministry are being raised (Hocking 1999, 2007, Melissen 1999, Rana and Kurbalija 2007).

Given that diplomacy is a key institution of the Westphalian state order (Held et al. 1999:39), perceptions of a transformation of the role of the state and eventually of the premises that the order as a whole is based upon have led a number of observers to suggest that a 'revolution in diplomatic affairs' is inevitable. IT is said to be a central factor in such a transformative dynamics. As one of the most influential reports

[4] This does not, however, mean that states would be losing their ability to exercise their sovereignty. As Rosenau (1999:349–350) notes, even though NGOs and advocacy groups were given access to negotiation sessions at such multilateral venues in the mid-1990s as the human rights summit in Vienna or the social summit in Copenhagen, when negotiations moved to their final stages, NGOs were not granted access and final agreements were reached by representatives of states. While this was the case also in the so called Ottawa process leading to the adoption of an international treaty banning landmines, states have recognized that in partnership with transnationally active NGOs they can achieve greater understanding and acceptance of their foreign policy agenda and get others to support their foreign policy goals (Price 1998, Axworthy 2003, Cooper 2004).

addressing this topic, the 1998 study entitled *Reinventing Diplomacy in the Information Age*,[5] stated

> "[t]he world is changing fundamentally. Images and information respect neither time nor borders. Hierarchy is giving way to networking. Openness is crowding out secrecy and exclusivity. [...] In this world of instantaneous information, traditional diplomacy struggles to sustain its relevance. [...] The prime mover of change is *information technology*" (p. 11, italics in the original).

This dynamic is said to have profound effects upon foreign ministries worldwide. As the British diplomat Steward Eldon (1994) argued in a pioneering study,

> "foreign ministries will have little option but to take advantage of IT if they are to remain competitive with other parts of government, their analogues in other countries, and in some respects, the media" (Eldon 1994:22).

Thinking along the same lines, the then Canadian Deputy Minister of Foreign Affairs, Gordon Smith (1997), announced the advent of *virtual diplomacy* and described the change dynamics in the following manner:

> There was a time when establishing a new embassy or diplomatic post took weeks, even months. Now, it takes a plane ticket, a lap top and a dial tone. And maybe a diplomatic passport. We can hit the ground running: this has huge implications for the mobility of our operations and what I call "just in time and place" operational effectiveness.

The broad availability of instant electronic information on national and international events coupled with the ease of international travel had supposedly "permanently altered the balance of power between foreign services and their political masters" (Langhorne and Wallace 1999:17). These effects of the IT-enabled global accessibility of informational sources are most pronounced when we look at the situation of diplomats in conflict areas, who are often less informed than their home government and/or dependent on the same sources of information such as the global news networks or web-sites. Hence, diplomatic reporting 'on the ground' from conflict-zones like post-2001 Afghanistan often remains a symbolic activity performed from behind the walls

[5] *Reinventing Diplomacy in the Information Age* Washington, DC: Center for Strategic and International Studies, November 1998.

of heavily guarded embassy compounds with information gathered from web-sites or at best through the narrow bullet-proof windows of armored military vehicles (Ross 2007:42–45).

Numerous reflections by academics addressed these developments focusing on the systemic impacts and the rise of new forms of diplomacy. Cooper (1998) observed how the *cyber-diplomacy* of Greenpeace undermined the legitimacy of France's nuclear testing in the Pacific. Ronfeldt and Arquilla (1999) proposed that *noopolitik* would be a new paradigm for diplomacy in which the gist of diplomatic activity would be shifting towards the creation and maintenance of soft power due to the growing importance for the relations between states of the technologically supported 'global realm of mind'—the noosphere. Metzl (2001) suggested there is a paradigm shift towards *network diplomacy* which involves a wide array of actors in shifting networks of cooperation that coordinate the foreign policy of a state. Reflecting about the role of U.S. embassies in 2015, Schmitz (2001) argued the following:

> Embassies were set up to be the most important intermediaries between one government and another. IT, however, disintermediates, and early 21st Century IT made the disintermediation of [government to government] relations both possible and desirable. [...] Embassies provide human, on-the-spot eyes and ears. Remote sensing, imaging teleconferencing, etc. will have eliminated much of that need. [...] Embassies 'show the flag' and otherwise 'represent' the sending state. IT has provided better and more immediate means for foreigners to understand the sending state.

Moses and Knutsen (2001) went as far as to suggest that the foreign ministry was no longer needed and coordination of diplomatic representation abroad could be performed by other agencies of the central state administration. Overall, the set of changes anticipated in the literature on the transformative effects of the technology driven dynamics are so deep and far-reaching that they amount to a possible revolution in diplomatic affairs in which diplomacy as we have known it for centuries is being transformed. As Dizard (2001:187) held,

> projections of a digital future increasingly set the tone and direction of U.S. diplomacy. [...T]heir impact cuts across every other foreign policy issue from nuclear proliferation to the environment. The result will be a different type of diplomacy, both in the problems it deals with and in the way it is organized.

Yet, more moderate voices, such as for instance those of Berridge (1995), Hocking (1999, 2001, 2007), Melissen (1999), Potter (2002) and Cooper (2004) had argued that diplomacy has a long history of adaptation and change, and the current IT-revolution meets resilient institutional structures and age-old habits (Bátora and Neumann 2002).

1.2 *Institutional Resilience of Diplomacy*

Diplomacy as a social activity involving mediation between human collectives has been around in various forms throughout history (Numelin 1950, Der Derian 1987, Hamilton and Langhorne 1995, Berridge 1995).[6] As Hocking (2001) points out, while

> it is true that diplomacy is an institution of the [Westphalian system] phase in the evolution of international relations, [...] it is equally true that diplomacy has a history that transcends the development of the nation-state and that it reflects a fundamental response, in the words of one observer, to a common problem of living separately and wanting to do so, while wanting to conduct relations with others.

Diplomacy in its modern form emerged along with the emergence of sovereign states in Europe between the 15th and 18th Century.[7] Since

[6] Since ancient times, persons directly involved in communicating with other collectives had to depart from their own social context and enter the social context of *the other*. The act of transgression of social contexts provided the persons performing it with special symbolic power both in the foreign context into which they arrived as foreigners and indeed also in their own domestic context upon their return. Discussing social exchanges among communities and cultures in early Bronze Age Europe around 2000 B.C., Kristiansen (2004) points out, that experience with foreign contexts provided travelers, traders and soldiers with high respect in their home community (sometimes up to the point of being perceived as having magical qualities) upon their return from the realm of the other, the unknown. They often strengthened and maintained these perceptions by carrying foreign items with strong symbolic value such as special kinds of swords or clothing which they acquired while "abroad".

[7] Prior to the rise and gradual institutionalization of what came to be called the Westphalian state order between the 15th and 18th Centuries, diplomacy was organized in a different way than today. It operated through multiple sets of norms and rules, and appeared sporadically whenever there was a need for communication between a myriad of actors involved in exchanges of various kind. The ancient Chinese empire, for instance, had no centralized hierarchical apparatus for conducting foreign relations, and instead featured provincial leaders, who were in charge of relations with neighbouring provinces (Hamilton and Langhorne 1995:8). In ancient Greece, there was no or only severely limited secrecy related to diplomatic dealings as the mandates of departing diplomatic missions were debated publicly at length in the home constituency and diplomatic messages were orated by diplomats openly in parliamentary assemblies abroad (ibid., pp. 9–10). And indeed, European states as late as in the 17th Century featured a plethora of diplomatic and quasi-diplomatic agents with no particular hierarchy among them (Anderson 1993). This was a situation when "all sorts of principals sent diplomatic agents to all sorts of recipients" (Queller 1967:11).

then, diplomacy has been a structure enabling regularized interaction between states and became a key institution of the interstate order. As such it has been anchored in the procedures, rules, routines, norms and professional culture of foreign ministries, which in aggregate represent the global organizational basis of diplomacy. Diplomats became the primary guardians of national interests and of information pertaining to the relations with foreign states. Specific diplomatic procedures, structures, routines and norms have emerged to that effect. Some of these are codified in the 1961 Vienna Convention,[8] while others are maintained as unwritten habits and codes of conduct of the transnational diplomatic community.

Throughout recent centuries, diplomacy has shown a great degree of resilience and adaptability (Berridge 1995, Hocking 1999, Melissen 1999, 2005). As Garrett Mattingly (1955:17) observed,

> a good many early fifteenth-century rules, procedures and types of documents survived the disintegration of medieval Europe almost unchanged and may still be recognized in contemporary diplomatic practice, surprisingly little altered by the passage of five hundred years. During the transition from medieval to modern times, in diplomacy as in some other fields, formal institutions changed less than might have been expected. It was the objects of policy and the vision of society which changed.

The diplomatic practices and routines have been embedded in the organizational structures and procedures of foreign ministries. Although today's foreign ministries are far larger and more complex than their organizational predecessors in previous centuries, they are still organized around geographical and functional divisions and are hence in principle not too different from foreign ministries in 17th Century Europe.[9] The right to conduct diplomacy and withhold foreign policy information remains an exclusive prerogative of the state, which endows foreign ministries of today with much of the same power and authority as they have had for centuries.

---

[8] According to the Vienna Convention of 1961, the main functions of a diplomatic mission include representation; protection of national interests; negotiation; information gathering and reporting; and promotion of friendly relations in various societal spheres.

[9] The French foreign ministry founded in 1626 by Richelieu featured geographical (territorial) divisions, divisions for consular affairs, administrative divisions and a cipher-bureau for communication. Given France's dominant position on the Continent in the 18th Century, the French foreign ministry served as a model that other countries establishing foreign ministries sought to emulate. See chapter 3 for more details about the development and institutionalization of foreign affairs apparatuses in Europe.

And yet, the alleged information technology revolution that in the eyes of some observers has been creating transparency and openness about international events on a scale previously unseen and leading to a radical decrease in the amount of foreign policy information exclusively guarded and exchanged by diplomats seems to strike at the very heart of the authority and power status that foreign ministries have been enjoying. The main research question that this study seeks to answer is therefore the following:

What are the effects of IT on the organizational basis of diplomacy?

The question is directed at investigating the nature of the IT-driven change dynamics. First, are the changes as radical and as extensive as the proponents of the IT-revolution in diplomacy claim and is there hence indeed a new kind of diplomacy emerging? If so, what are its characteristics? Or do we find evidence to support the moderate views which point to the institutional resilience of diplomacy and which also suggest that changes are slow, gradual and path-dependent? In other words, what is the *magnitude of change*? Second, are the responses to the technological change dynamics consistent in all foreign ministries around the globe or are there differences in terms of how various national foreign ministries adapt? Are the change dynamics coherent across the spectrum of foreign ministries or are there differences in adaptation? In short, what is the *direction of change*? Overall, an assessment is sought made of whether the technology-driven dynamics transforms the organizational basis of diplomacy and its structures as we know them, or whether the organizational basis of diplomacy yet again proves resilient and adapts in a path-dependent manner.

1.3 *IT-revolution or Institutional Resilience?*

The by now fairly large volume of contributions addressing various aspects of IT-driven change in diplomacy is marked by a myriad of interesting observations, reflections, and descriptions of the processes that are taking place at foreign ministries around the globe. Yet as Hocking (2001) points out,

> [o]ne of the problems confronting any serious evaluation of the changing nature and role of diplomacy is [...] the lack of analytical as opposed to descriptive material that surrounds it. This is reflected in the largely unprofitable arguments regarding the 'decline' of diplomacy [...]

The analytical lacuna which Hocking points at concerns both the understanding of diplomacy, as well as the understanding of the ongoing set of processes that might amount to an IT-revolution in diplomatic affairs. If indeed such a revolution is taking place, the existing institutional structures of diplomacy would have to be replaced by a new set of structures that foster new kinds of power and authority relations.

There are two separate bodies of literature on diplomacy, from which insights relevant to the topic of the current study can be drawn. On the one hand, the *diplomatist literature* focuses on various aspects of diplomacy as they emerged throughout history, on the negotiation process, on diplomatic functions, on the diplomatic professional culture, habits, practices, routines and their historical roots.[10] These are important contributions about various aspects of diplomacy as a social and political activity. Virtually all of the authors allude to the fact that modern diplomacy has been a resilient and adaptive set of institutional structures which has been changing rather incrementally in recent centuries. Yet none of the studies has used insights from new institutionalist theory in organizational analysis consistently, which would allow them to *conceptualize diplomacy as an institution* and establish the core organizing principles of 'modern' (Westphalian) diplomacy to get a better sense of both magnitude and direction of current change.[11] Such a theoretical

[10] Examples include works by Satow (1922), Nicolson (1939, 1953), Numelin (1950), Der Derian (1987), Anderson (1993), Berridge (1995), Hamilton and Langhorne (1995), Hocking (1999), Melissen (1999), Berridge (1995), Neumann (2001). For an elaborate overview of contributions to the diplomatist literature see Sharp (1999).

[11] To be sure, there are contributions addressing various aspects of the institutional or systemic character of diplomacy. This includes the work of authors of the English School in international relations such as Herbert Butterfield, Martin Wight, Hedley Bull and Adam Watson (for an insightful discussion of these authors' views on diplomacy see Neumann 2003). Furthermore, a number of more recent contributions by IR scholars have alluded to institutional and socially reproduced characteristics of diplomacy (see for instance Ruggie 1993, Spruyt 1994, Wendt 1998, Reus-Smit 1999, Holsti 2004). Using the approach of historical sociology, Jönsson and Hall (2005) also seek to establish the essential and timeless features of diplomacy as an institution, or as they put it, the "common denominators characterizing diplomacy across time and space" (p. 24). Yet none of the contributions mentioned here uses new institutionalist approaches in organizational analysis consistently to conceptualize diplomacy as an institution featuring an organizational basis structured in the interplay with prevailing political orders in specific historical periods, i.e. currently the 'modern' (Westphalian) state order. This involves the roles of diplomats as well as organizing standards maintained through isomorphic pressures among foreign ministries constituting the global organizational field of diplomacy. These concepts are elaborated in chapter 2, where I address the processes that led to the structuration of diplomacy as an institution of the modern state order.

conceptualization is missing also in the second strand of the relevant literature, namely in what might be called the *IT-revolutionary literature on diplomacy* which focuses on various aspects of the technology-driven change dynamics in the international environment perceived as challenging diplomacy.[12] Again, most of the studies in this strand of the literature point at important tendencies and make interesting observations. Yet given the lack of a clear understanding as to what exactly are the core organizing principles of diplomacy that would have to be transformed if diplomacy were to be declared transformed, conclusions announcing new kinds of diplomacy emerging or its decline are anecdotic, partial and sketchy (albeit often good-sounding and shrewd). In sum, an overview of the two strands of relevant literature on diplomacy suggests that neither provides us with coherent analytical tools to allow us to properly understand the changing nature of diplomacy through the impact of IT.

The first analytical goal here is therefore to conceptualize diplomacy as an institution. In doing so, the new institutionalist perspective in political science associated with March and Olsen (1984, 1989, 1995) is conducive and will be used. It provides us with an understanding of institutions as sets of rules and routines that define appropriate actions in terms of roles and situations. A second strand of new institutionalist literature that will be used here includes contributions addressing the dynamics of distribution and maintenance of particular organizational forms and practices across a set of organizations (e.g. the foreign ministries). This approach is epitomized by the notion of the *organizational field* in the sense of DiMaggio and Powell [1991]). As will be shown in Chapter 2, these two approaches from organization theory are complementary and highly relevant to the conceptualization of diplomacy as an institution of the interstate system and its embeddedness in organizational arrangements at foreign ministries. Armed with a clearer understanding of the character of diplomacy as an institution and its core organizing principles, we can get a better analytical grip on the IT-driven change dynamics it is subject to.

In exploring IT-effects from an institutional perspective it is also necessary to understand how IT is embedded in social structures (i.e. Bijker et al. 1987, Lyon 1988, Orlikowski 1992, Castells 1996). Institutional dynamics may be a driving force behind the revolutionary impact of

[12] Examples of contributions are in part 1.1.

IT,[13] but they may also mediate and limit effects of IT.[14] This duality forms the two sides of a nexus between IT-revolution and institutional resilience that the current study seeks to explore in the context of diplomacy. Two scenarios of IT-driven change dynamics in diplomacy can hence be anticipated.[15]

First, it is the *transformation scenario* (IT revolution) which implies a replacement of the core organizing principles of diplomacy by a new set of organizing principles; changes at individual foreign ministries would conform to homogenizing pressures of the IT-revolution and there would be no major differences between individual foreign ministries in applying IT in support of their operations. Secondly, it is a *status-quo scenario* (institutional resilience), which would imply that the effects of IT would leave the core organizing principles of diplomacy in place; change at individual foreign ministries would be guided by internal institutional dynamics reflecting organizational histories and cultures. While these two scenarios represent outermost positions on the analytical continuum of the magnitude of change, effects might be mixed, and a third model could therefore be helpful as a middle position—the *renewal scenario*.[16]

To test the assumptions of the transformation scenario which implies a convergence of effects and a drive towards homogenization around a new set of IT-supported structures and practices of diplomacy that are different from the established structures and practices, empirical evidence is needed from foreign ministries with varying institutional backgrounds and varying levels of IT-implementation. Obviously, such

---

[13] One way to conceptualize the social embeddedness of IT is to see it as a form of an organizational field in the sense of DiMaggio and Powell (1983). As such it is constituted by the producers, users, tech-support staffers and consulting gurus, and distributes homogenizing pressures leading to convergence in IT-supported structures and practices across organizational contexts. Catalyzed by a sense of performance crisis (see March and Simon [1958], Cyert and March [1963]) or by able change management by an IT-savvy leadership (see Buchanan and Boddy 1992, Hammer and Champy 1993, Andersen 1999) such a drive towards IT-supported homogenization of structures and practices at foreign ministries might lead to replacement of existing institutional structures and procedures of diplomacy by a new set of structures. This would effectively amount to a revolution.

[14] As March and Olsen (1989) point out, institutions are carriers of stability and institutional change is seen as gradual and path-dependent involving historical inefficiencies and local equilibriums. According to this approach, effects of IT will be mediated by the established institutional setting of diplomacy and there will be active resistance to change.

[15] The scenarios are inspired by the hypotheses in Olsen and Sverdrup (1997).

[16] The models are developed in more detail in Chapter 3.

empirical evidence will also serve to test assumptions of the status-quo scenario, where differences between foreign ministries in terms of structures, practices and procedures are assumed to remain in place even with technological tools in place. The choice of the foreign ministries of Canada, Norway and Slovakia as cases to be investigated in the current study was hence made owing to their varying institutional histories and administrative cultures, and indeed to their different levels of IT-implementation.

Canada's Department of Foreign Affairs and International Trade (DFAIT)[17] has been a world-leader among foreign ministries in implementing innovative IT-solutions in support of its operations on a global scale (Potter 2001). Due to Canada's somewhat ambiguous emergence as a sovereign state, DFAIT has been embedded in the context of the domestic political-administrative system to a larger extent than has likely been the case in most other foreign ministries (Eayrs 1982, Burchill 1993). The *Royal Norwegian Ministry of Foreign Affairs* (N-MFA), on the other hand, has been keeping up a traditional role of a foreign ministry with its strongly formalistic style and a level of detachment from the rest of the governmental administration (Neumann 1998, Neumann and Leira 2005). It has been severely hesitant in taking IT into use (Bátora and Neumann 2002). Finally, the *Ministry of Foreign Affairs of the Slovak Republic* (MFA SR) is a new foreign ministry of a new state founded in 1993. Both its organizational structure and it technological infrastructure have been under turbulent development since its foundation (Bátora 2000).

To sum up, the main contributions of this study could be summarized along the following lines:

- Neither the diplomatist literature, nor the 'IT-revolutionary literature on diplomacy' provides us with an analytical framework for assessing the impacts of IT-effects on the organizational basis of diplomacy. This study conceptualizes diplomacy as an institution embedded within a specific set of institutionalized organizational structures of foreign ministries and develops a framework for examining its change.

---

[17] By a late 2003 decision of Prime Minister Paul Martin, DFAIT was split into Foreign Affairs Canada (FAC) and International Trade Canada (ITCan). The two Departments were again amalgamated in 2006. Moreover, as the data for this study was collected in 2001–2003, most of the findings presented here pertain to DFAIT, and this acronym is hence used throughout the study.

- While in-depth knowledge of IT-driven change processes at foreign ministries is needed to assess the changing nature of the organizational basis of diplomacy, it is puzzling indeed that there are still virtually no case-studies of IT-effects on foreign ministries. This study is a first step toward filling this lacuna as it provides three comprehensive case studies of IT-effects on the foreign ministries of Canada, Norway and Slovakia.
- Exploring IT-related change in the organizational basis of diplomacy might provide at least partial answers to the question raised by Coolsaet (1999:3) as to whether we are "witnessing an adjustment of the state system today or a real 'time break', a systemic transformation comparable to the transition of the medieval world order to the modern state system?" As Hocking (2007:4) argues, studying change in these organizations helps us to appreciate significant shifts in the international system and the ways how states adapt to such shifts. Hence a focus on the effects of the IT-revolution in foreign ministries in the current analysis.

1.4 *Organization of Chapters*

With the aim of conceptualizing diplomacy as an institution, the following chapter first addresses the historical processes that were key to the structuration of diplomacy as an area of political activity. It focuses on the gradual institutionalization of sovereignty as a constitutive logic of the state system and introduces core organizing principles of interstate diplomacy that co-evolved with the rise of sovereign states. Relevant new institutionalist approaches in political science are then introduced and applied to conceptualize the processes that led to the rise of diplomacy as an institution of the interstate order. Core features of the organizational basis of modern diplomacy are identified. A general model is developed and nuances are added by a brief discussion of variations in how sovereignty is embedded in states.

Chapter 3 then outlines an analytical framework for studying the effects of IT in the context of foreign ministries. First, the chapter briefly addresses the introduction of the telegraph in the mid-19th Century as the first revolution in information and communications technology and its impacts in foreign ministries. This is followed by a discussion of the current information revolution and its impacts on the operational conditions of public administrations in general and of foreign ministries more specifically. Theoretical notions of IT as a socially embedded technology are then introduced and modes of IT-supported institutional change are discussed. Based on the insights from Chapter 2, operationalizations of core organizing principles of diplomacy are then elaborated, and these are then used in the development of three scenarios of IT-effects on the organizational basis of diplomacy—transformation, status-quo and renewal. Research methodology and the choice of the foreign ministries of Canada, Norway and Slovakia as case studies are then discussed. Conclusions follow.

Chapters 4, 5 and 6 are empirical chapters summarizing the findings in respectively DFAIT, N-MFA and MFASR. Each chapter first introduces the respective foreign ministry by describing its institutional background and characteristics. This is broken down into three elements—organizational structure, personnel and domestic environment. A brief description of the technological infrastructure of the respective organization follows. Effects of IT are then described in relation to the three organizing principles of diplomacy identified in Chapter 2 and their operationalizations developed in Chapter 3. In the conclusion to each of the empirical chapters, an overall assessment of the case is provided relating the findings to the scenarios of IT-effects on diplomacy developed in Chapter 3.

Chapter 7, the concluding chapter, first summarizes the theoretical and analytical framework. The main findings on IT-effects on the core organizing principles of diplomacy as well as other relevant observations and findings are then outlined providing a basis for a comparative assessment of change dynamics in the three cases. The main drivers of change are then addressed, which is followed by a set of reflections on the theoretical lessons of the current study. In the concluding part, the possibilities of an IT-supported revolution in diplomacy are discussed along with potential avenues of future research.

CHAPTER TWO

THE ORGANIZATIONAL BASIS OF MODERN DIPLOMACY: ITS EMERGENCE, CHARACTERISTICS AND CHANGE

2.1 *Introduction*

Throughout recent centuries, diplomacy has proved to be a resilient and adaptive set of practices (Nicolson 1939, 1953, Mattingly 1955, Berridge 1995, Hocking 1999, Melissen 1999). With the rise and gradual structuration of what came to be called the Westphalian system of states, diplomacy was embedded within states and became the organizing institution of the modern state order (Bull 1977, Watson 1982, Neumann 2002, Bátora 2003, 2005, Jonsson and Hall 2005). Today, one is struck by the compatibility of arrangements for organizing diplomacy in states around the world, despite differences in political regimes, administrative traditions, geopolitical specifics and national interests.[1] Likewise, one also notes the shared sense of diplomatic community providing diplomats from various states with a shared professional identity. Drawing upon insights from new institutionalist approaches in political science and organizational analysis, the goal of this chapter is to outline the processes that led to the institutionalization of diplomacy in its modern form and identify the features that characterize its modern organizational basis consisting of the global aggregate of foreign ministries.

Although examples from history are used, it is not a historical account. Inspired by the approach of Tilly (1975), the rationale here is not to present a comprehensive account of the historical developments guiding the emergence of diplomacy,[2] but to shed light on specific historical processes that led to the emergence and institutionalization of the core organizing principles of the diplomatic system that still characterize it today. The historical accounts are hence merely used as a tool for

[1] See Steiner (1982), Hocking (1999), Cascone (2000) for examples of how foreign affairs administrations around the world are organized.

[2] Others have done so. See for instance Andreson (1993), Hamilton and Langhorne (1995).

highlighting the constitutive elements of diplomacy as an institution and their emergence in specific historical circumstances, which will enable us to better grasp their current robustness and the leeway for their adaptability and change. Devising such a conceptualization—or an archetype in Peters' (1999:104–105) sense—will pave the way for the development of the analytical framework for investigating the magnitude and the direction of change in diplomacy due to effects of IT presented in the following chapter. As Peters (ibid.) argues,

> development of archetypes of institutional forms [is useful] for comparative purposes. The logic here is similar to that employed by Weber in his development of 'ideal type' methodology [...]. The variation in forms of organizations and institutions is sufficiently great that any attempt to examine them all would have the researcher bogged down in almost endless detail. Therefore, is appears more efficient to create ideal types of institutions against which to compare the institutions observed in the real world.
>
> As well as being useful for comparative purposes, the archetype analysis is useful for speaking to questions of change in institutions. [...O]ne of the dangers of focusing on institutions for analysis is that they tend to be relatively permanent, and even inflexible, so that change is difficult to detect. The argument accompanying use of archetypes is that institutions can only change from one archetype to another. [....] The movement from one archetype to the other involves then a process of deinstitutionalization and a subsequent reinstitutionalization, as one set of structures is replaced by the new alternative.

In other words, developing an archetype conceptualization of the modern organizational basis of diplomacy is necessary to establish whether an IT-driven revolution in diplomacy is taking place.[3] In what follows, this chapter first traces the processes that led to the gradual embedding of diplomacy within states. The core organizing principles of modern diplomacy are then identified and a general model of a foreign ministry as a carrier of these principles is introduced. In the second part of the chapter, new institutionalist theoretical approaches in organizational analysis are used to conceptualize the character of diplomacy as a system or structures and norms regularizing interstate interaction. Conclusions follow.

[3] See the definition of revolution in footnote 2 in part 1.1.

2.2 *The 'Informational Prerogative' of the State and the Organizing Principles of Modern Diplomacy*

Diplomacy has always been about communication and mediation between human collectives. An essential precondition for relations between human collectives is at least a basic level of adherence to certain shared norms and rules enabling these relations. In other words, as soon as receiving the messenger and using him/her for delivering a message back to the sender took precedence over killing the messenger, an elementary form of diplomacy was taking place (Hamilton and Langhorne 1995:7). Settling conflict or divergence in interests by means of communication and negotiation have since ancient times been the most essential attributes of diplomacy (Berridge 1995, Cohen and Westbrook 2000). Prior to the rise and gradual institutionalization of what came to be called the Westphalian state order between the 15th and 18th Centuries, diplomacy was organized in a different way than today. It operated through multiple sets of norms and rules, and appeared sporadically whenever there was a need for communication between a myriad of actors involved in exchanges of various kinds. The ancient Chinese empire, for instance, had no centralized apparatus for conducting foreign relations, and instead featured provincial leaders, who were in charge of relations with neighbouring provinces (Hamilton and Langhorne 1995:8). In ancient Greece, there was no or only severely limited secrecy related to diplomatic dealings as the mandates of departing diplomatic missions were debated publicly at length in the home constituency and diplomatic messages were orated by diplomats openly in popular gatherings and assemblies abroad (ibid., pp. 9–10, see also Watson 1992:47–68). And indeed, early medieval Europe was an environment in which "all sorts of principals sent diplomatic agents to all sorts of recipients" (Queller 1967:11). Negotiations were conducted by various kinds of envoys. They were called *legatus* or *nuncius* or referred to descriptively for instance as "our faithful man" (ibid., p. 4). The open character of the diplomatic function is apparent in a 13th Century definition of a legate:

> A legate is, or can be called, whoever has been sent from another [...] either from a ruler or from the pope to others [...] or from any city or province to a ruler or to another... or even from a proconsul. [...] On this account a legate is called a substitute for the office of another. [...] But also nuncii whom foreigners send to us are called legates. (quoted in ibid. p. 6)

Moreover, an 'embassy' had a temporary character and its mandate could last all from delivering a message to a representation for several months or even years.[4] The early medieval As Mattingly (1955:27) points out,

> [i]n the thirteenth and fourteenth centuries, not only princes and free cities of the empire, and the greater feudal nobles, but even merchant towns, even universities and craft guilds, sent formal quasi-diplomatic agents on occasion, apparently without anyone's questioning their right to do so, or finding it odd to refer to them as ambassadors (*legati*), while the legists continued to discuss diplomatic agents under the same rubrics which dealt with all agents legally empowered to act for others.

This, from today's point of view, fragmented and disorganized myriad of diplomatic platforms and forms operated in an environment marked by common Christian traditions, values and norms. Indeed, the people of Europe in the 13th and 14th Centuries, divided as they were in terms of their allegiance to various kinds of kings, dukes, barons, knights and other principals saw themselves as being part of a common Christian Commonwealth, a *res publica Christiana*, embodied by the papacy and building upon traditions of the Roman empire (ibid. pp. 18–20). Emergence of consensus about diplomatic rules, norms and procedures, and about the rights and status of diplomatic envoys was hence facilitated by the shared normative basis of all actors engaged in diplomatic communication.[5] As Stinchcombe (1965:149) following Weber argues, "universalistic religions making oaths to strangers sacred, universalistic law making contracts between strangers binding, reliable, negotiable instruments so that one can trust the paper and not the man [...] clearly make it easier to construct social systems out of groups of strangers." Adherence to universalistic shared norms by all actors did not, however, preclude a gradual stratification of the emerging diplomatic system. Power differences started to play a role

[4] Until today this historical experience remains codified in particular language expressions. For example the German word "Botschaft" means 'a message' but also signifies 'an embassy'.

[5] Referring to the 1436 'Short Treatise About Ambassadors' by Bernard du Rosier, the archbishop of Toulouse, Mattingly (1955:45) specifies that "[a]mong all peoples, in all kingdoms and lands, [ambassadors] are guaranteed complete freedom in access, transit and regress, and perfect safety from any hindrance and violence. [...] Those who injure ambassadors, or imprison them, or rob them, who impede their passage, or even abet or approve such acts are properly regarded as enemies of mankind, worthy of universal execration. For whoever interferes with ambassadors in their public function injures the peace and tranquility of all."

in terms of asserting ones authority over a particular realm, and kings of France, Spain or England would often refuse to receive envoys from smaller feudal lords on equal terms.[6] What is more, they would start to regard the reception of envoys from their own subjects by other kings as a suspicious and sometimes even a hostile act (Mattingly 1955:27). Hence, although late medieval diplomacy was in theory still a system with a relatively low threshold for entry and participation, in day-to-day practice and through enactment by powerful actors a pattern started to emerge, where only fully sovereign principals would be regarded legitimate participants.[7] This development was related to what March and Olsen (1995:4–5) refer to as a "great metamorphosis in thinking about politics", where the previous conception of political order as divine and natural was substituted by a new conception of political order as secular and based on human will, calculation and power. As Huntington (1966:389) argues,

> the most significant difference between modern man and traditional man is in their outlook on man in relation to his environment. In traditional society man accepts his natural and social environment as given. What is ever will be: it is or must be divinely sanctioned; and to attempt to change the permanent and unchanging order of the universe and society is both blasphemous and impossible. Change is absent or imperceptible in traditional society because men cannot conceive of its existence. Modernity begins when men develop a sense of their own competence, when they begin to think that they can understand nature and society and can then control and change nature and society for their own purposes. Above all, modernization means the rejection of external restraints on men, the Promethean liberation of man from control by gods, fate, and destiny.

This ideational shift had numerous implications for the way governance was organized within the newly formed sovereign units. The sovereignty

---

[6] Status differences played a role even between kings. As late as 1558, Russian Tsar Ivan IV would write to Erik XIV of Sweden that "the (Holy) Roman Emperor and other great sovereigns are our brothers, but it is impossible to call you a brother because the Swedish land is lower in honor than those states" (quoted in Anderson 1993:20).

[7] Ideas of sovereignty had their origin in earlier political thought. For instance, according to the 14th Century Perugia lawyer Bartolus of Sassoferato, only those principals and rulers were sovereign who acknowledged no superior authority and were hence "princes in their own domains" (quoted in Mattingly 1955:27). This rule obviously applied mostly to kings and city states, and indeed to the Papacy. According to Buzan, *sovereignty* is "the claim to be the ultimate political authority, subject to no higher power as regards the making and enforcing of political decisions." (Barry Buzan "sovereignty" *The Concise Oxford Dictionary of Politics*. Ed. Iain McLean and Alistair McMillan. Oxford University Press, 2003. *Oxford Reference Online*. Oxford University Press).

principle was first realized in the city states of the 15th Century Italy. Independent from both the emperor and the Papacy, Milan, Florence, Naples, Venice, and some others, had fairly well organized governance structures and were able to control their respective territories from one center to an extent which was not yet thinkable in larger states at that time. In addition they were able to concentrate quite a high number of well educated men to serve as their ambassadors, and their foreign policies compared with policies of larger states were better organized and more continuous. The city of Milan is considered to be the first to have had *resident ambassadors* in other northern Italian cities as well as abroad at the court of Sigismund, the king of Hungary and the Holy Roman Emperor elect in the years 1425 to 1432 (Mattingly 1955:71). In an effort to preserve their sovereignty under continuous threats of invasion by larger European states, Italian city states started to appoint resident ambassadors to courts in France, Spain and to the Habsburgs in the second half of the 15th Century. Spain was the first to establish a resident embassy in Venice in 1495, and the Habsburgs, France and England eventually followed suit in establishing permanent embassies in Italy and in each others' capitals. By the end of the 16th Century, the originally northern Italian system of resident embassies was well on the way of becoming the dominant European diplomatic practice (Hamilton and Langhorne 1995:38).

In larger European kingdoms, claims of sovereignty were legitimized by the late 16th Century theory of *the divine right of kings*, which as Huntington (1966:385) observes, "admirably served the purposes of the modernizing monarchs of the seventeenth century by giving the sanction of the Almighty to the purposes of the mighty." Societies within these emerging state-formations were increasingly being penetrated by the central authority of the state (see for instance Tilly 1975, Rokkan 1975).[8] What is of relevance in the context of the current study is that this ideational development, which was empowering self-interested sovereigns by building upon pre-modern universalistic notions

[8] These processes had often a violent character and led to wars, where the to be sovereigns had to fight contesters within the territory over which they claimed final authority. It was symptomatic that Richelieu had the defense walls of French cities demolished in the 1640s so as to disable resistance against central authority of the king, while Peter the Great of Russia went as far as publicly hanging hundreds of *boyars*, who resisted his centralization efforts in the late 17th Century (Holsti 2004:32–33).

of order embedded in God's will, and which hence was a "transition stage between medieval and modern politics" (Huntington 1966:385), legitimized sovereign actors' refusal to interact on equal terms with other than sovereign actors. Sovereignty hence gradually started to be regarded as a *sine qua non* condition for legitimate participation in diplomatic exchanges, and recognition as a sovereign actor by other sovereign actors was crucial for an actor's ability to engage in diplomatic relations (Ringmar 1996). Diplomacy hence increasingly came to be regarded as the realm of sovereign actors (Anderson 1993, Hamilton and Langhorne 1995, Holsti 2004).

Claims of sovereignty in the conduct of foreign affairs implied claims of the right to send and receive ambassadors. By claiming this right, sovereigns effectively asserted the right to issue authorized (or official) foreign policy instructions on behalf of their state, and indeed also the right to receive authorized (official) foreign policy information from abroad. The sovereign (the state) hence acquired what might be termed an *informational prerogative in foreign affairs*—the right to create, receive and store authoritative (or official) information on the foreign affairs of the state.[9] While actors other than the sovereigns continued to conduct relations with foreign counterparts, with the gradual institutionalization and diffusion of the sovereignty principle, the foreign affairs informational prerogative of sovereigns was strengthened and the information exchanged between actors other than sovereigns was accordingly degraded in status. If then, as Holsti (2004:185) points out, "[n]orms surrounding the institution of diplomacy were very much tangled up with the issue of sovereignty and the status of dynasts", one might argue that a central element in that 'entanglement' was the claim of the informational prerogative in foreign affairs by the sovereign (the state). In other words, the implication of the increasing acceptance of the norm of sovereignty was that a distinction was being made between what was considered official and authorized foreign affairs information exchanged between sovereigns (states), and what was considered other

[9] According to Jonathan Bradbury, "prerogative powers are those which are at the autonomous disposal of heads of state and which do not require sanction by a legislature. Their theoretical justification lies in Locke's view of a need for a final arbiter to maintain order." Jonathan Bradbury "prerogative" *The Concise Oxford Dictionary of Politics*. Ed. Iain McLean and Alistair McMillan. Oxford University Press, 2003. *Oxford Reference Online*. Oxford University Press.

kinds of foreign policy information.[10] In a sense, the sovereigns (states) and their diplomatic representatives hence did *not* become *gatekeepers* in relations between their realms and other sovereign political entities—an imagery often used to describe the role of the state agencies in charge of foreign affairs—but rather *authorizers* of information flows between home and abroad.

To manage the centralized flow of information pertaining to foreign affairs of the state, specialized groups of dedicated officials had to be created at the respective sovereign courts, and the foundations for the nuclei of future foreign ministries were hence laid. The French foreign ministry founded by Richelieu in 1626 is of particular importance here as it developed an unprecedented level of administrative sophistication and became a widely imitated model of foreign affairs administration. Under Louis XIV (1643–1715) and his foreign minister Jean Baptiste Colbert, the ministry developed standardized procedures for recordkeeping, document registers, and started to develop geographical specializations. As Lauren (1976:8) observes, "[i]n order to develop specialized expertise, they created two separate geographical divisions to manage all matters of external relations. The first handled Spain, Portugal, Britain, the German states, and the Papacy, while the other administered affairs with Turkey, the Italian states, Poland, Denmark, Russia, Sweden, and Switzerland. Administrative specialization also developed with the gradual emergence of distinct internal services. These included the *Cabinet du Ministre* (Minister's Cabinet), the Cypher Bureau, and the *Bureau des Fonds* (Accounting Office) charged with managing the ministerial bureaucracy, issuing passports, conducting surveillance on foreigners, handling matters of protocol, and supervising appointments and transfers." As Anderson (1993:79) argues, 18th Century France has hence developed a "well-organized [foreign] ministry of a recognizably modern sort", and as Hamilton and Langhorne (1995:74) propose, France's foreign ministry "had reached a stage of development by 1789 which others were only to achieve in the nineteenth century". Indeed, France hence became what Huntington (1966:412) calls the "pattern state" which other states on the Continent and beyond sought to copy in terms of organizing the governmental and indeed

[10] It is this difference in status of information exchanged, which is at the heart of the distinction between *high politics* and *low politics* often promoted by diplomats and often denied by other actors.

also the diplomatic machinery. Through such isomorphic processes, the foreign ministry as a type of organization had become (and as Hocking (2007) points out, until today ideal-typically remains) a filter in the exchange of official messages between the domestic and the foreign environment; a repository of skills and advice on foreign negotiations; and not least an information bank on foreign affairs of the state.[11]

The ultimate goal of asserting and maintaining the sovereign's (the state's) informational prerogative in exchanges with foreign powers had implications for the way the conduct of the newly emerging state-diplomacy was organized. This entailed *hierarchy* enabling the generation of unified official foreign policy standpoints and their unified representation; *secrecy*, enabling the drawing of boundaries between official foreign policy information which could be kept secret and other foreign policy information; and finally, following the rise in importance of public opinion and public participation of citizens, *one-way communication with the public* enabling the sovereign (the state) to publicly share information to the extent and at a time when this was deemed appropriate.

2.2.1 *Hierarchy*

Following the rise of sovereignty as the constitutive logic of the political order in Europe, diplomats representing a sovereign as his or her direct impersonations were bestowed with the same divine authority as the sovereign, and ranked hence hierarchically above all other officials in the service of the sovereign (Anderson 1993, Hamilton and Langhorne 1995). This shift entailed that the status of the ambassador increasingly became an object of interest of the aristocratic circles and a diplomatic career gradually became the domain of aristocracy. The dominance of aristocrats meant that issues of ceremony and detailed attention to status differences stemming from diplomatic hierarchy became commonplace in diplomatic dealings throughout European capitals. As an anonymous late 17th Century Spanish guide to diplomatic procedure states,

> [a]n ambassador should be specially careful to maintain the authority of his position and assert his prerogatives, as for instance not giving the door or the right hand to any minister [diplomatic official lower in the

[11] The role of the foreign ministry as a centralized information bank on the state's foreign affairs is also crucial for governmental decision-making in the current environment of complex interdependence. See note 1.

> diplomatic hierarchy than the ambassador] or individual, whether of his own country or another. [...] I refer to the ceremony of offering the door, the chair or the right-hand seat in the coach (cf. Anderson 1993:58).

Official written diplomatic communication also illustrates the level of attention given to differences related to hierarchical ranks. As Harold Nicolson ([1939]1988:106) instructs his readers:

> The words used in termination [of a letter] vary according to the rank of the person addressed. To an ambassador, the Secretary of state signs as follows:
>
> I am, with great truth and respect,
> Sir,
> Your obedient servant,
>
> [...] To a minister, the termination is identical except that the word 'regard' is submitted for the word 'respect'. To a chargé d affaires, both the respect and the regard are omitted and the line ends (somewhat curtly perhaps) with the word 'truth'.

The social composition of diplomatic services manned mostly by aristocrats well into the 20th Century[12] was a factor strengthening the focus on hierarchy as a central norm at foreign ministries. The prevalence of aristocrats meant that issues of ceremony and detailed attention to status differences stemming from diplomatic hierarchy became commonplace in diplomatic dealings throughout European capitals. Processes of bureaucratization at the end of the 19th Century further strengthened the hierarchical culture at foreign ministries. This involved implementation of elements of bureaucracy as described by Weber (1978:218–220)—clearly defined hierarchy of offices; clearly defined spheres of officials' competence; impersonal official obligations; fixed salaries; separation of private from public property of the means of production or administration; written files; selection based upon competence; specialized training; the office as the sole career for the incumbent; and not least promotion dependent upon judgment of superiors. These developments were accompanied by a broader and incremental process, which Rosenberg (1958:111) referred to as the *reorientation* of the aristocratic community of governmental administrators. This involved a "transformation of group outlook and habits of living. [...which made them more likely] to develop an occasional interest in long hours and

[12] To take one of many examples, the German foreign service consisted of eight princes, twenty-nine counts, twenty barons, fifty-four untitled noblemen, and only eleven commoners as late as 1914 (Lauren 1976:27).

hard and efficient work, and [gave] their avocational activities a more vocational complexion" (ibid.). Through these broader bureaucratization processes, established professional norms and practices of the aristocratic diplomatic community were gradually complemented by a new set of structures that came to characterize national governmental bureaucracies at large. As a late 19th Century German contemporary observed, at the Ministry of Foreign Affairs,

> [t]he strictest order prevails from top to bottom, unconditioned obedience is the rule, and, as is right and proper, everyone obeys without protest or contradiction, whatever his own opinion may be. Everything downstairs moves at the bidding of one absolute subordination. [...] There must be no stoppage caused by this or that individuality. Acquiescence is the first and highest law (Busch 1898, quoted in Lauren 1976:31).

Hierarchy hence no longer was just a professional norm inherent in the foreign service, but through bureaucratization it became embedded in the structure, rules and standard operating procedures of foreign ministries. Lines of authority and communication were standardized and foreign ministries could to a greater extent than before act as unified organizational actors. Moreover, the growing complexity and volume of relations that had to be handled meant also that an increasing number of decisions were made by various officials in the bureaucratic line, even middle-ranking ones, because "it was no longer possible for almost every question to go to the minister himself for final decision, as had hitherto been normal" (Anderson 1993:118). With the introduction of telegraph, embassies were to a greater extent integrated to the overall bureaucratic machinery of foreign ministries (Jones 1983:116–138). Throughout this process the once fairly extensive freedom of action of ambassadors in distant capitals became infringed and foreign policy decision-making became more centralized.[13] The comment by the experienced British diplomat, Sir Francis Bertie, who after spending four decades working in the Foreign Office was about to be sent as resident ambassador to Rome in 1903, indicates the increasing level of centralization and bureaucratization in the relations between headquarters and embassies: "In Downing Street [the Foreign Office] one can at least pull the wires, while an ambassador is only a d-d marionette" (quoted in Hamilton

[13] Yet, Stratford Canning, the 19th Century British ambassador to Constantinople, argued otherwise, namely that due to the brevity of messages and instructions received via telegraph, the resident ambassador had in fact multiple ways of interpreting them and taking action accordingly (Hamilton and Langhorne 1995:132).

and Langhorne 1995:132–133). Hence, throughout the process of bureaucratization which went hand in hand with the communications advances, the power in decision-making has definitely shifted towards the headquarters, where it was increasingly embedded in bureaucratic routines and practices.

Even in the early 21st Century, strict hierarchy remains a characteristic in most foreign ministries. As Ross (2007:107) reports on the practices in the British FCO, "if a junior diplomat writes the first draft, it must be checked by a senior diplomat before being 'signed off'. Particularly important messages must be checked by ambassadors themselves, since it is their name that goes at the end of the message (itself an unconscious reinforcement of the hierarchicalism of the system)." Similar observations regarding the practices in the Norwegian foreign ministry were reported by Bátora and Neumann (2002).

To sum up, hierarchy remains an organizing principle enabling foreign ministries to produce unified negotiation positions and unified foreign policy statements in particular situations and thereby live up to the expectations related to their role as the ultimate source of authoritative information on the foreign affairs of the respective state.

2.2.2 *Secrecy*

A ubiquitous aspect of diplomatic negotiations whether conducted by *ad hoc* envoys or permanent ambassadors has been the need to protect the information exchanged from parties other than those involved in the actual negotiation. Since the Middle Ages, the most common way to protect diplomatic communication was the use of cyphers. The earliest record of such practices is a Venetian diplomatic document from 1411, but it was the city state of Milan—a pioneer in establishing permanent embassies—that played a pioneering role in the development of cypher techniques.[14] In the late 17th Century, the French foreign ministry was the first to establish a Cypher Bureau (Lauren 1976:8). This had since become a standard practice and virtually all foreign ministries of today feature organizational units ensuring secure diplomatic communications. Another early practice used to protect diplomatic communication was

[14] As Anderson (1993:22) reports, the *Nationalbibliothek* in Vienna possesses no fewer than 200 cypher keys used by the Milanese government between the years 1450–1496.

the use of diplomatic couriers, who remain a constant feature of most foreign ministries world-wide. The norms of secrecy and the institutionalized procedures applied to put it into practice effectively enable the diplomats to protect and thereby maintain the informational prerogative of the state in foreign affairs.

Besides the instrumental reasons for secrecy in diplomatic dealings, normative reasons for it had emerged with the gradual formation of the professional diplomatic community in the 17th and 18th Centuries. According to some of the central theorists of diplomacy of the time, such as Abraham de Wicquefort and later François de Callières,[15] it was a question of professional dignity for diplomats to be honest in their dealings. For that to be possible, diplomatic negotiations were to be kept secret so as to enable the generation of a highest possible amount of trust between the parties involved, because trust was a necessary condition for the successful accommodation of the often competing interests between the rulers or governments that the diplomats represented (Hamilton and Langhorne 1995:69).[16] Secrecy hence became and remains a central norm informing the diplomatic professional community.

Through the process of bureaucratization, application of secrecy was formalized in rules and procedures of foreign ministries. As Weber (1978:225) points out, secrecy is an inherent feature of bureaucracy:

> [b]ureaucratic administration means fundamentally domination through knowledge. This is the feature which makes it specifically rational. This consists on the one hand of technical knowledge which, by itself, is sufficient to ensure it a position of extraordinary power. But in addition to this, bureaucratic organizations, or the holders of power who make use of them, have the tendency to increase their power still further by the knowledge growing out of experience in the service. For they acquire through the conduct of office a special knowledge of facts and have available a store of documentary material peculiar to themselves. While not peculiar to bureaucratic organizations, the concept of 'official secrets' is

[15] Abraham de Wicquefort (1680–81): *L'Ambassadeur et ses Fonctions*, 2 volumes, The Hague; François de Callières (1716): *De la manière de nègocier avec les souverains, de l'utilité des négotiations, du choix des ambassadeurs et des envoyez, et des qualitez nécessaires pour réussir dans ces employs.*

[16] This openness within the diplomatic community and a parallel wariness in relations to both the domestic society and foreign publics had implications for the institutionalization of the norm of secrecy as an integral part of the diplomatic mindset. As Thomas Paine observed in 1792, "[t]he diplomatic character is of itself the narrowest sphere of society that man can act in. It forbids intercourse by the reciprocity of suspicion; and a diplomatic is a sort of unconnected atom, continually repelling and repelled" (quoted in Neumann 2001a:610).

> certainly typical of them. It stands in relation to technical knowledge in somewhat the same position as commercial secrets do to technological training. It is a product of the striving for power.

Through the process of bureaucratization, matters of official foreign policy were formalized as the specific area of technical expertise that foreign ministries were to administer. The foreign ministries of today are the repositories and guardians of the 'knowledge growing out of experience', i.e. a store of special knowledge related to details of international treaties and other aspects of foreign relations of the respective state assembled in records of internal archives, to which the public continues to have limited access. Bureaucratic procedures and routines, such as the *need-to-know principle*,[17] ensure the maintenance of secrecy. This is fostered also through what in line with the Equipped for the Future report (1998) might be called a 'culture of secrecy' common in foreign ministries and featuring a host of specific formal and informal routines. Ross' (2007:91) observations based on his Foreign Office experience capture the essence of the processes at the level of individual diplomats and are hence worth quoting at some length:

> The signature of the Official Secrets Act marks one initiation into the culture of secrecy that pervades government, and particularly those parts dealing with foreign policy. When you learn how to handle documents, for instance, you are taught that the originator of the document must classify it, using designations starting with 'restricted' up to 'top secret'. You are taught that only those documents that would not perturb you if they were handed out to passers-by on the street can be designated 'unclassified'. Unsurprisingly therefore, almost every document produced inside the Foreign Office is classified 'restricted' or above. This culture is constantly reinforced throughout one's career. Telegrams are transmitted only when highly encrypted. All computers are hardened against electronic eavesdropping. Telephones carry stickers warning against divulging state confidences. So many and so ubiquitous are these limitations that it is soon clear that the only people (with whom) one can discuss candidly what 'we' are doing are one's colleagues—other members of the club of 'we'. For what 'we' are doing is the affairs of state, and other states might try to find out our secrets; therefore one should only talk to people with a 'need to know'. This excludes almost everyone, including those in whose name 'we' are acting.

[17] The *need to know principle* regulates access to information at foreign ministries by allowing foreign service officers to access only the documents and files of their respective organizational unit.

In sum, secrecy embededded in specific procedures and rules reinforces internal hierarchy in the foreign ministry, as classified information is only conveyed to a few selected officers usually occupying higher positions in the system, as well as places the foreign ministry (along with a few other agencies) at the top of information access hierarchy within the government.[18] Secrecy is hence an essential element in the maintenance of the informational prerogative of the state in foreign affairs.

2.2.3 *One-way communication with the public*

The existence of a professional diplomatic community implies different modes of communication between actors *within* the community and in relation to actors *outside* of it, i.e. the public. As indicated above, communication within the diplomatic community has a regulated character following rules, norms and procedures of emerging diplomatic practice and ceremony. Those rules manifest themselves in the particular diplomatic language that came to distinguish communication within the diplomatic community from communication with non-diplomatic actors. As Nicolson ([1939]1988:122–123) notes,

> the need of intelligence [in relations between governments] is self-evident, but the equally vital need of tact is often disregarded. It is this latter need that has led diplomatists to adopt a paper currency of conventionalised phrases in place of the hard coins of ordinary human converse. These phrases, affable though they may appear, possess a known currency value.

Furthermore, communication among diplomats came to be marked by the general norm of reciprocity and exchange of information, essentially a form of *dialogue* mapping out the positions of counterparts and gathering of relevant information, which preceded actual foreign policy decisions. As Watson (1982:17–18) observes,

> [t]he European diplomacy which our global system inherited developed as a dialogue between members of a system which (as always in the past) had a cultural and historical identity strong enough to ensure that its members recognized certain rules.

[18] As Ross (2007:74) observes, "there is a tendency in government to see intelligence material as being at the pinnacle of the hierarchy of information. Unlike the voluminous flow of diplomatic telegrams, memos and open source information that hits computers on desks across government every day, intelligence arrives in slim folders, adorned with colorful stickers announcing not only the secrecy of the information therein, but the restricted circulation it enjoys. The impression thus given, a product of these aesthetics, is of access to the real thing, the secret core denied to all but the elite few."

Communication within the diplomatic community is hence rule-based, dialogical and happens often *ex ante* relative to foreign policy decisions that eventually ensue.

A fairly different set of patterns emerged in relation to actors outside of the diplomatic community. In the early days of modern diplomacy between the 15th and the 18th Centuries, the subjects of sovereigns were in general not deemed worthy of being informed about issues of foreign policy, informing the public about foreign policy as we would understand it today was hence not taking place. Yet, what could be considered a basic form of informing the public about foreign relations were the entry ceremonies of new ambassadors. Political and diplomatic messages of various kinds were conveyed to the gathered public (including present foreigners) during such ceremonies. The 1691 return of William III of England to the Netherlands is a case in point. As Jones (1984:13–14) points out:

> [upon William's return] Dutch officials could think of no more fitting gesture than a triumphal entry into The Hague, complete with appropriate allegorical figures. Peace, War, Justice, Tyranny, and Felicity adorned the triumphal arches that had been raised for William's procession through the city [...]. Then, lest the English who had accompanied William on his return to the country of his birth should miss the points that were being expressed through symbolism, the Dutch sponsors of the triumph put their messages clearly in Latin. The inscriptions on the same arch read: 'To the Pious, Happy, Renowned William the Third, the Triumphant Father of his Countrey, Governor, Stadtholder, and Restorer of the United Netherlands. England's Liberator, Scotland's Preserver, Ireland's Pacificator, now returned.' And again: 'After great things done at home and abroad, as having made a strict League with the Princes, the Revenger of the Subjects' wrongs and Defender of the oppressed.'
>
> The pageantry was entirely political, and the messages were diplomatic. The whole tangled history of the succession of the English throne, the Revolt of the Dutch United Provinces against Spanish rule in the Netherlands, and the Grand Alliance of England, Holland and the League of Augsburg against the expansionist France under Louis XIV could be read in the arches, banners and illuminations that signaled William III's triumphal entry into The Hague.

Besides messages on foreign policy, entry ceremonies of ambassadors could communicate a great deal about the wealth and power of the sending sovereign as the arriving ambassador would put gold chains, the strength of his suite and bodyguards, the number of horses in his carriage, and the like on flamboyant public display (Anderson 1993:16).

Following the invention of printing press, the growth of literacy and the ideational developments of the Enlightenment, public opinion—i.e. all the non-governmental opinions which were given public expression in the press, pamphlets, provincial assemblies, the universities, salons and other societies of intellectual and political elites (Hamilton and Langhorne 1996:124)—started to play a role that could no longer be ignored by the diplomatic services. Metternich noted in 1808, that "public opinion is the most powerful medium of all. Like religion it penetrates into the darkest corners" (quoted in ibid.), while lord Palmerston would note some decades later that "[o]pinions are stronger than armies. Opinions, if they are founded in truth and justice, will in the end prevail against the bayonets of infantry, the fire of artillery and the charges of cavalry" (cf. Nicolson [1939]1988:37).

Efforts to shape the public opinion were first introduced on a larger scale during the reign of Napoleon I. The French 'Grande Armeé' published a daily bulletin which had a noteworthy influence on the perception of events by the peoples of the occupied territories, Napoleon's diplomats founded a newspaper in Poland for propagandistic purposes, and the French government published an English-language newspaper 'Argus' to foster anti-British propaganda (Anderson 1993:137). Throughout the first decades of the 19th Century, it became standard practice for ambassadors to publish articles explaining policies of their government in the most respected newspapers in the country of their residence. What could be called *a press release mode of communication* remains a standard form of communication with the public applied by foreign ministries in their management of public communication around the world until today.

The spread of literacy did not only lead to public demands for information on foreign policy but also for information on the foreign ministries and diplomatic services themselves. Foreign ministries hence started to publish information on their own organization and staff. The first step in this direction was taken in Sweden, where they began to publish an annual list of its ambassadors and consuls as early as 1824. Great Britain, France and Italy followed suit in the 1850s and 1860s (Hamilton and Langhorne 1995:127). Around the same time, some foreign ministries also started to publish earlier diplomatic correspondence pertaining to particular foreign policy matters. The British Foreign Service published in the 1820s and 1830s the so called 'Blue Books' which included selected diplomatic correspondence on selected foreign policy events from earlier years, with other European foreign services gradually engaging in similar publishing practices—the French had their "Yellow Books", while Bismarck's foreign ministry would publish its "White Books" (Hamilton and Langhorne 1995:127). As Anderson (1993:114) observes, "[p]ublications of this sort broke sharply with the anonymity and complete immunity from public scrutiny which had hitherto marked foreign offices." To deal with the more broad and unstructured demands for information on foreign policy by journalists, the business community and the general public, *press departments* were created at most foreign ministries in the period around World War I. Their role was to draft press releases, organize press conferences, and generally provide information on foreign policies to both the domestic and foreign audiences.[19] Public communication at foreign ministries has hence traditionally been centralized and channelled through particular units established for that purpose. The traditional norm in foreign ministries' communication with the general public has hence been that of a *one-way provision of information* (press release format) in an *ex-post* fashion.[20] Conforming to the notion of the state's informational prerogative in foreign affairs, foreign ministries of today continue to see it as legitimate to release information on official foreign policies when and to the extent they deem appropriate.

In sum, foreign affairs administrations of all states reflect the organizing principles of hierarchy, secrecy and one-way public communication (see Table 1 below), which enables each of the states participating in the international system to maintain its informational prerogative. The consequent ability of every state to issue, receive and store authorized information on relations with other states represents a systemic feature enabling states to communicate in a regularized manner. Diplomacy can hence indeed be considered a system of communication between states in the sense of Deutsch (1960, 1978) or Wight (1978) or an interstate dialogue in the sense of Watson (1982). What could explain the diffusion and maintenance of such a system of rules and procedures conforming to a particular set of guiding organizing principles throughout the world

[19] For examples of practices in the French and German foreign ministries in the first decades of the 20th Century see Lauren (1976:111–117, 147–152).

[20] Informing about foreign policy decisions and actions made, but not involving the public in ex ante discussions or consultations of foreign policy priorities and planned actions.

Table 1: Foreign ministry as an organizational carrier of diplomacy

Embedded in: Organizing principles of diplomacy	*Structures and practices of foreign ministries*	*Norms and role-expectations in the diplomatic community*
Hierarchy	– Hierarchical line relations, bureaucratic divisions – hierarchy between HQ and embassies	– hierarchy of diplomatic ranks – diplomats as top mediators of their state in relation to other states
Secrecy	– organizational units in charge of secure communication (e.g. cypher bureau) – procedures for handling of information (e.g. the *need to know* principle)	– secrecy as a norm enabling honesty in diplomatic exchanges – matters of official foreign policy can be subject to secrecy
One-way Public Communication	– centralized units for public communication (e.g. press department, spokesperson)	– the norm of one-way communication – the norm of ex-post communication

in the absence of an overarching authority? Addressing this question, the next section uses new institutionalist theoretical approaches to present a view of diplomacy as an institution embedded within and among states.

2.3 *Diplomacy as an Institution of the Interstate Order*

Given the nature of diplomacy as an organized system of rules and procedures informing relations between states in a more or less stable manner in changing political circumstances over the last two to three hundred years, it is of relevance in the context of the current study to draw upon insights of new institutionalist approaches in political science and organizational analysis.[21] The following sections first outline some

[21] Scott (2001) traces the beginnings of institutionalism in Europe to the works of Karl Marx, Emile Durkheim and Max Weber. While these institutionalist contributions

of the core theoretical notions and then, using these notions, a conceptualization of diplomacy as an institution of the interstate system is developed.

2.3.1 *Institutions and institutionalized action*

Institutional orders are created at certain points in history, when actions of individuals become regularized, habitualized and standardized through the gradual development of impersonal rules of conduct and stable behavioural patterns connecting roles and situations. Over time, such interactions between actors lead to the emergence of what Durkheim has called *social institutions*, which "are a product of joint activity and association, the effect of which is to 'fix', to 'institute' outside us certain initially subjective and individual ways of acting and judging" (Alexander 1983, cf Scott 2001:13). A set of standard operating procedures is established and individuals in a given collective become "used to it" sometimes to the extent that these procedures and rules are taken for granted. Throughout the institutionalisation process a shared set of normative principles guiding behavior is developed, which is anchored in the so called *logic of appropriateness* (March and Olsen 1989, March and Olsen 2006, Olsen 2008). Before taking action, individuals do not always rationally consider all available options in order to maximize their self-interest, but also follow routines and habits. An action sequence of an individual socialized into a set of institutionalised practices could be modelled in the following manner: "What kind of a situation is this? Who am I? How appropriate are different actions for me in this situation? Do what is most appropriate" (ibid., p. 23). Institutions can hence be defined as *"collections of interrelated rules and routines that define appropriate actions in terms of relations between roles and situations"* (ibid., p. 160). Institutions are not perceived as something exogenous by individuals, because through interaction with other individuals in specific institutional contexts they gradually internalize a given set of practices,

are often referred to in more contemporary organizational analyses, they were first and foremost works in sociology and political philosophy. The beginning of the institutionalist approach in organizational analysis can be traced back to the works of Philip Selznick (1949, 1957). For useful overviews of the various strands of new institutionalist theory see March and Olsen (1989), DiMaggio and Powell (1991), Peters (1999), Scott (2001, 2004). For views on the emerging avenues of new institutinoalist theory see March and Olsen (2005).

they acquire "the institutional lenses". Action is then seen as a way of fulfilling the obligations related to a particular social role or identity (March and Olsen 2004:1). Hence,

> [a]ppropriateness refers to a match of behavior to a situation. The match may be based on experience, expert knowledge, or intuition, in which case it is often called 'recognition' to emphasize the cognitive process of pairing problem-solving action correctly to a problem situation. [...] The match may also carry with it a connotation of essence, so that appropriate attitudes, behaviors, feelings, or preferences for a citizen, official, or farmer—essential not in the instrumental sense of being necessary to perform a task or socially expected, nor in the sense of being an arbitrary definitional convention, but in the sense of that without which one cannot claim to be a proper citizen, official, or farmer (March and Olsen 1995:30–31).

As observed by Olsen (1992), institutional dynamics can be explored by focusing on the points of interplay and tension between individuals, structures and environments. He refers to these as *micro-*, *meso-* and *macro level* of institutions focusing respectively on a) power and rational decisions of individual actors purposefully choosing strategies that shape structures; b) structural arrangements and internal dynamics within an institution that guide processes of change and institutional reactions to respectively changes in the environments and individual initiatives, and finally c) contextual factors in the institutional environment that determine the existing institutional structures through selection processes (ibid., pp. 248–249). The interplay between individuals, structures and environment(s) is operationalized in organized activities and structures. Most institutions are represented in the society through various kinds of organizations that form each institution's organizational basis.

2.3.2 *Organizational fields as carriers of institutional logics of action*

The process of institutionalization involves development of notions of *exemplary behavior* in the sense of Olsen (1992:252). These notions indicate what an actor of a particular kind is expected to do in a particular situation, i.e. they embody the logic of appropriateness. Organizational arrangements and sanction mechanisms are developed around the maintenance of patterns of exemplary behavior in a particular sphere of society, which hence lead to the establishment of *organizational fields* (DiMaggio and Powell 1983, Meyer and Scott 1983). One of the most usual definitions of an organizational field suggests that it is a collection of organizations,

> that in the aggregate constitute a recognized area of institutional life: key suppliers, resource and product consumers, regulatory agencies, and other organizations that produce similar services and products (DiMaggio and Powell [1983]1991:64–65).

An organizational field emerges through the process of *structuration* (in the sense of Giddens), which involves the patterning of social activities and relations through time and across space (Scott 2001:75). According to DiMaggio and Powell (1991:65), structuration has four elements: increased interaction between organizations within the field, emergence of clearly defined patterns of domination and coalition building, an increased information flow within the field, and the emergence of mutual awareness of participating organizations of their involvement in a shared enterprise. Organizational fields serve hence as transmitters of organizational practices and structures between participating organizations, and hence maintain shared meanings and thereby facilitate transactions at the inter-organizational level within the field, and of the field as a whole with other spheres of social activity. At the same time they serve as transmitters of standards, which form the basis of actors' legitimacy and status. New actors entering the field need to adopt established practices and structures within the field in order to gain legitimacy, access and acceptance as members of the organizational field. Hence, organizational fields support isomorphic change in participating organizations.

DiMaggio and Powell (ibid., pp. 67–74) identified three sources of isomorphic institutional change: coercion, mimesis and normative pressures. *Coercive isomorphism* stems usually from "formal and informal pressures exerted upon organizations by other organizations upon which they are dependent and by cultural expectations in the society within which organizations function". A dominant power position enables an actor or a set of actors to exert coercive pressures on less powerful actors, who in the struggle to achieve legitimacy opt for isomorphic changes.

Actors opt for isomorphic change also in the absence of coercive pressures. In attempts to reduce uncertainty about their own situation, actors often imitate and mimic established practices and organizational structures, which they perceive as efficient and legitimate. This practice can be referred to as *mimetic isomorphism*. If one knows the predominant organizational arrangements in a given organizational field, one is able to predict the later organizational make-up of the newly emerged organizations entering the field, due to the likelihood of mimetic isomorphism.

Finally, isomorphism can also be the result of professional standards governing an organizational field, which can easily determine which practices and behaviors are acceptable and legitimate within that field. Professional associations and guilds guarding professional jurisdictions within a field yield incentives for *normative isomorphism* among participating actors.

The theoretical ideas sketched here will be used in section 2.3.4 below to conceptualize diplomacy as an institution embedded, on the one hand, in a system of transnationally shared meanings and professional codes of conduct providing the basis of the diplomats' logic of appropriateness, and, on the other hand, in a shared system of compatible organizational arrangements at foreign ministries building the basis of the organizational field of diplomacy. Before moving on to this conceptualization, one more theoretical suggestion is in order, namely that the notions of organizational field and logic of appropriateness convey different and complementary understandings of institutionalized action. Addressing this complementarity is in its place, for it will provide us with a more complex understanding of the inherent dynamic of change and stability in diplomacy.

2.3.3 *The two-dimensional character of institutions*

As pointed out earlier, actions of an actor following the logic of appropriateness are based on a deep appreciation of his/her own identity and role in a given situation (March and Olsen 1989, 1995, 2004). These are situations in which actors act according to their innermost *conviction* about their stance in particular circumstances.[22]

The notion of organizational field inspires a different imagery of action. Here, the actors seek legitimacy for their actions by adopting what are seen as legitimate structures, procedures, language etc. in an isomorphic manner. The action sequence of an actor within an organizational field could be modelled as follows: 1) What am I member

[22] *Conviction* may be defined as "An attitude dimension which is concerned mainly with how a person is predisposed to think about a situation" (Oxford Reference Online, Oxford University Press (available at http://www.oxfordreference.com/views/ENTRY.html?subview=Main&entry=t161.e1548)

of? (Who is like me?)[23] 2) How is this situation perceived by others who are like me? 3) Do what others who are like me do in a situation like this. Appropriateness in an organizational field is hence established not by inner soul-searching of actors/organizations but rather by external monitoring of the collective patterns within the field and by consequent isomorphic adoption of established *conventions*.[24] Behavior within an organizational field hence does not necessarily involve matching of actions with internalized identities of actors, but rather involves a matching of actions with collectively accepted standards—conventions—within the field.

Conventions within a field evolve gradually in interactions among a multitude of actors in the process of field structuration, but they are usually stabilized by the most powerful (resourceful) actors (DiMaggio 1991, Parkhurst Fergusson 1998). Once established, they serve to decrease complexity by providing legitimizing templates for appropriate organizing, appropriate procedures and actions. Organizations within a field hence adopt conventions of organizing and other features irrespective of their actual effectiveness (Meyer and Rowan ([1977]1991), Tolbert and Zucker (1983), DiMaggio and Powell ([1983]1991). Through organizational structures and procedures standardized within a field, *appropriateness is routinized*, as actors adopting the field-wide standards of organizing and conduct, do not need to get involved in complicated processes of soul-searching and matching identities to actions to decide what actions would be appropriate in which situations. Actors follow collectively adopted routines and practices embedded in standardized organizational arrangements, which enables them a claim of acting appropriately.

As institutions are present in our world through arrangements organizing social and political life including shared meanings, roles, norms, rules and formal organizations, both kinds of institutionalized action addressed above, i.e. action informed by the logic of appropriateness (convictions) and action informed by the collective rationality

---

[23] This first question was formulated to me by Walter Powell in a private conversation in reaction to my suggestion that the first question in a model of an action sequence in an organizational field should be "Where am I?".

[24] *Convention* may be defined as "a way in which something is usually done, esp. within a particular area or activity. [...] To attract the best patrons the movie houses had to ape the conventions and the standards of theatres." Oxford Reference Online, Oxford University Press (available at http://www.oxfordreference.com/views/ENTRY.html?entry=t140.e16440&srn=3&ssid=1004147023#FIRSTHIT)

of organizational fields (conventions), occur often simultaneously. This is consistent with March and Olsen's (2004:4) recent definition of an institution as

> a relatively stable collection of rules and practices, embedded in structures of *resources* that make action possible—organizational, financial and staff capabilities, and structures of *meaning* that explain and justify behaviour—roles, identities and belongings, common purposes, and causal and normative beliefs. Institutions are organizational arrangements that link roles/identities, accounts of situations, resources and prescriptive rules and practices (italics in the original).

While structures of resources are subject to economic pressures and/or technological change, which may result in organizational change, structures of meaning (roles, identities) may be more durable and often remain more stable. On the other hand, roles and identities of actors/organizations may be changing, while structures of organizational resources may remain stable due to existing standardization within organizational fields. Institutions are hence kept stable by the stability of the roles and identities which they embody, by the rigidity of organizational arrangements which they are embedded within, or by the stability of both.

Yet, while institutions in this way bring about stability of social systems, this two-dimensional character of institutions leads also to intra-institutional tensions or collisions, which, as Friedland and Alford (1991) argue, are among the primary drivers of institutional change. Actors may for instance argue in favour of particular organizational change in order to maintain a particular kind of institutional identity in a changing environment. Such tensions are inherent in any institution, because, as Eisenstadt (1964:245–246) observes, institutional systems are never fully homogeneous and groups within the system will come up with their own specific interpretations of institutionalized sets of norms and rules of exemplary behavior, they will argue that other organizational arrangements should be put in place so as to maintain the norms and rules of the institutional system, and/or organizations carrying the norms and rules of an institutional order develop their own specific interests and needs. The very process of institutionalization hence plants the seeds of change. As Eisenstadt (ibid., p. 253) puts it,

> institutionalization of any social system creates possibilities for specific and defined types of change, which develop not randomly but in specific directions to a large extent set by the very process of institutionalization (ibid., p. 253).

The following part uses the theoretical notions discussed in the above to address the systemic character of diplomacy as an institution of the state order.

2.3.4 *Institutionalization of modern diplomacy*[25]

Modern diplomacy and territorial (later nation-) states co-evolved in a mutually constitutive set of processes, which makes diplomacy both carrier and product of the interstate system. Most social structures exhibit this dual role in that they are "both the medium and the outcome of the practices they recursively organize" (Giddens 1984:25, cf. Scott 2001:75). As pointed out by Der Derian (1987:106–107), it is no coincidence that modern diplomacy and the Westphalian state system evolved as mutually reinforcing concepts, because

> what uniquely characterizes the paradigm of diplomacy is its utility for states in balancing the forces of hegemony and anarchy. In other words, diplomacy emerges as the collective and reflexive embodiment of the states' ultimate task—self-preservation in an alien environment (ibid., p. 111).

In other words, diplomacy is both a function and a determinant of the international order (Hamilton and Langhorne 1995:238). A central precondition for the functioning of diplomacy as a system of norms and rules regulating interstate relations is the existence of a common institutional basis shared by all states. This was provided for by the rise of sovereignty as the core logic structuring the political landscape of medieval Europe and making the sovereign state the core unit of political organization. As representatives of states and exclusive guardians of the sovereigns' informational prerogatives in foreign affairs, diplomats started to develop particular norms, rules, routines and procedures pertaining to the conduct of their shared enterprise. These conditions created a favourable situation for institutionalization of diplomacy. As Watson (1982:17) points out,

> in the European society of states, diplomacy has emerged as an organizing institution, bearing its distinctive styles and manners and its own networks of procedures, rules, treaties and other commitments. The European system, so organized, was able to exercise assertiveness and

[25] This part builds upon Bátora (2005).

> restraint over their members because they were bound from the beginning by much more than mere political arrangement. [...] And it is generally recognized that the sophisticated techniques and heightened awareness of how the states system operated, which European diplomacy required from its independent member states, contributed not a little to the remarkable phenomenon, contrary to the experience of other states systems [e.g. Hellenistic, early Chinese and Indian], that no single state proved to be so powerful that it could for any length of time absorb or even dominate all the others.

The organizational field of diplomacy was *structured* through an increased interaction between foreign ministries of various European countries as the primary actors in 19th Century international relations. Stable patterns of coalition building and domination were clearly present in the European diplomatic system in particular after its standardization at the Congress of Vienna in 1815. The fact that foreign ministries were constituted primarily to administer the diplomatic archives growing in volume as more and more diplomatic dispatches began to flow into governmental offices from embassies and from foreign governments is clear evidence of an increase in information flow between the participating organizations within the field. And finally, a fundamental principle embedded within the diplomatic system is that of mutual recognition of diplomatic agents and their rights (such as immunity), which shows the growing awareness of actors within the field as being involved in a common enterprise. Over time and through regularized interactions, the conduct of diplomacy, as Anderson (1993:ix) points out, "came increasingly to be seen in terms of ideas and ideals which gave unity and some underlying intellectual structure [...] to the growing volume of diplomatic activity", so that diplomats from various countries have gradually developed a shared professional identity. As Nicolson ([1939]1988:14) points out:

> [b]y 1815 therefore the Diplomatic Services of the nations had been recognized as a distinct branch of the public service in each country. A definite profession had been established, possessing its own hierarchy and rules, and destined [...] to evolve its own freemasonry and conventions.

The Congress of Vienna in 1815, as Neumann (2002b) observed, was a turning point because specific diplomatic practices were codified and formalized, and because this was a form of diplomatic interaction that went beyond discussing particular treaties and situations. In other words, the Congress has set up the practices and routines for future relations between states. The common set of professional norms, rules and values have since been transmitted through standardized socialization procedures at foreign ministries so that professional diplomats around the globe form a group of professionals with

> the corporate feeling, which the diplomatic service creates. Even as scientists, philatelists and other experts find, when they meet together, that the interests of their calling transcend all differences of nationality or language, so also do the diplomatic services of the several countries evolve a form of solidarity and establish certain tacit standards which they all respect (Nicolson [1939]1988:40).

As such it is plausible to think that they structure their actions according to a particular logic of appropriateness anchored in diplomatic rules, norms and principles. Through training and socialization at foreign ministries, individual diplomats become carriers of the systemic logic. As Ross (2007:21) argues, "when you become a diplomat, you are encouraged to submit yourself to the collective state: your individual 'I' becomes 'we'." And he goes on to explain, that "by some subterranean and unexplained process when you join the Foreign Office, you begin to identify yourself with the state" (ibid., p. 85). In this way, the norms of diplomacy penetrate individual diplomats in the respective foreign ministry and diplomacy as a system enabling regularized interstate interaction can work.

Without the structuring by sovereignty as the constitutive principle of the interstate-order reflected in organizing principles of diplomacy and in organizational structures and norms of foreign ministries, interactions between political entities would have the form of what March and Olsen (1989:21) call an open structure of free political competition, in which it would not be clear "what would happen, or who will do what to whom when". With diplomacy in place, states as actors in the international arena do not act randomly but fulfill particular roles determined by the established set of diplomatic rules, norms and principles. Indeed, states' actions in the international arena only make sense within the established institutional framework of diplomacy—it designates who are legitimate actors and participants in the inter-state system, what are legitimate actions and situations. The institutionalized structure of diplomacy hence decreases complexity in interstate relations and thereby increases capability of states as actors in the international environment. In this way, diplomacy makes actions of states understandable and to some extent predictable due to the limited number of available (or legitimate) options. Despite differences in cultures, political regimes and interests, states share a common institutional identity to a large extent defined and determined by the institutional framework of diplomacy, so that, as Wendt (1999:242) points out,

> [t]he vast majority of states today see themselves as part of a 'society of states' whose norms they adhere to not because of on-going self interested calculations that it is good for them as individual states, but because they have internalised and identify with them.

Similarly, as Krasner (1999:184–185), referring to the English School in international relations theory, argues,

> actions [of states] follow particular patterns not because they are dictated by some higher authority or coerced by a threat of force, or constrained by the power of other states, but because players in international society have a common worldview. [...] The existence of international society is reflected in diplomatic practices.

Hence, to rephrase the original definition of institutions by March and Olsen (1989:160), *diplomacy as an institution is a set of rules and routines that define appropriate actions of states in the international environment in terms of relations between their roles as states and situations.* This set of rules and routines is informed by the constitutive logic of sovereignty, which entails the notion of the sovereign's informational prerogative in foreign affairs and forms the basis of the *logic of appropriateness* informing the role of the state as a state. It is embedded in and maintained by the core organizing principles of diplomacy—hierarchy, secrecy and one-way public communication—which carry the essence of modern diplomacy as an institution in the sense that without structures, procedures, norms and routines associated with hierarchy, secrecy and one-way communication with the public, a diplomat cannot claim to be a proper diplomat. These organizing principles are reflected in structures, procedures and norms at foreign ministries, which in aggregate form the organizational field of diplomacy (see Figure 1 below). Interactions of states are hence informed by a shared systemic logic of action emanating from the notion of sovereignty and by a shared set of organized practices, which decrease complexity and enable effective exchanges irrespective of cultural background and national-administrative traditions.

Obviously, this is a conceptual model and the way interstate diplomacy is organized is far more complex. First, it is clear that not only diplomats, but also heads of state, prime ministers or other actors conduct

Figure 1: Diplomacy as an institution of the inter-state order

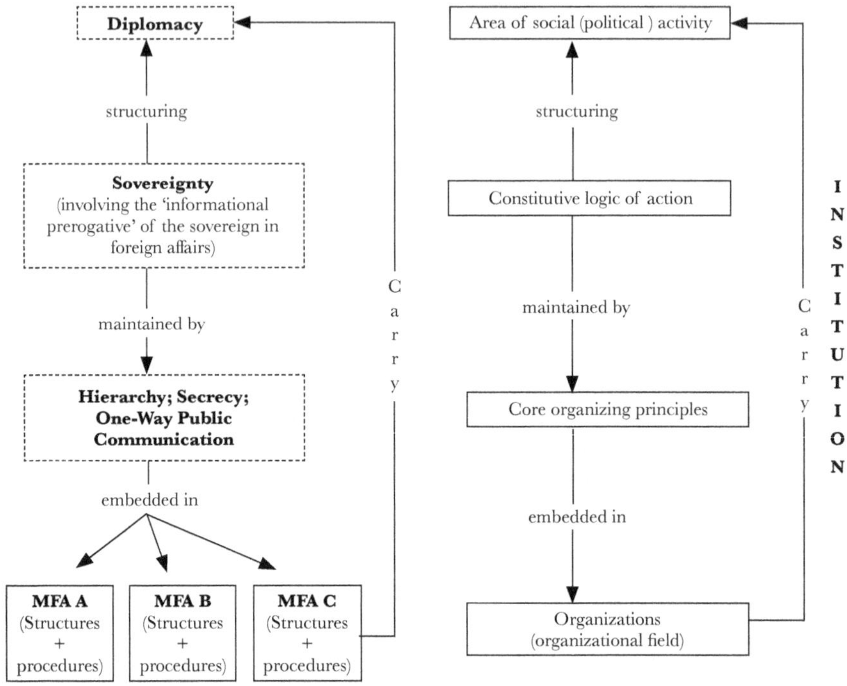

diplomatic relations on behalf of a state. Second, while sovereignty is the constitutive logic of the state system, differences in political traditions and political systems between states have led to variations in the embedding of sovereignty. As Huntington (1966) argues, concentration of sovereign authority in the centralized state was a specific of France and more broadly of the states on the European Continent such as Prussia, Sweden, or Spain. Although the development in Britain was similar, it led to different results. According to Huntington (ibid., p. 380),

> In Britain, too, church was subordinated to state, authority was centralized, sovereignty asserted internally and externally, legal and political institutions differentiated, bureaucracies expanded, and standing army created. The efforts of Stuarts, however, to rationalize authority along the lines of continental absolutism provoked a constitutional struggle, from which the Parliament eventually emerged the victor. In Britain, as on the Continent, authority was centralized but it was centralized in Parliament rather than in the Crown.

While in Britain, the all-powerful legislature became the counterpart of the all-powerful monarchs ruling the continental European states, sovereignty was embedded in yet a more peculiar way in the United States. There, political traditions of Tudor England became pervasive and the country

> continued to adhere to fundamental law as both a source of authority for human actions and an authoritative restraint on human behaviour. In addition, in America, human authority or sovereignty was never concentrated in a single institution or individual but instead remained dispersed throughout society as a whole and among many organs of the body politic. Traditional patterns of authority were thus decisively broken and replaced in Europe; in America they were reshaped and supplemented but not fundamentally altered. The continued supremacy of law was mated to the decisive rejection of sovereignty (ibid., p. 382).

Although the role of the state hence apparently varies internally in different countries around the world, it remains *constant externally* in relation to the system of interstate relations embodied in diplomacy, where the sovereign state conducts authorized relations to other such entities. As Nettl (1968:564) points out, the international function of the state

> is an invariant; countries with a low degree of 'stateness' in the intrasocietal field have to make special differentiated provisions accordingly (like the special status of the British Foreign Office vis-à-vis other governmental bureaucratic organizations [...], and the very special status of foreign affairs in federal societies like the United States and Switzerland, where they are one of the primary raisons d être for the claim of stateness on the part of the federal government). This difference between extrasocietal and intrasocietal problems of control and action is much less marked in countries with a strong state tradition like Germany and France. Hence [...] even where the notion of the state is very weak, as in Britain and the United States, the effective extrasocietal or international role is not affected.

Expectations about the role of the state as the carrier of centralized authority developed in the absolutist states of 17th Century continental Europe (notably France) form the basis of the logic of appropriateness carried in the diplomatic system, i.e. the logic of action epitomized in the notion of *raison d'État*. No matter how distant such a logic may be for the political constituencies of countries with different notions of the state, they are expected to act according to the logic of the system. This leaves foreign ministries of countries like Canada or the U.S. struggling to perform the role of a unifying top coordinator of foreign affairs that their counterparts from countries with the European continental tradition of statehood perform with much greater ease. Simplifying somewhat, while foreign ministries in countries with a continental

European tradition play the role of the *Ivory tower* both domestically and internationally, foreign ministries in countries like Canada and the U.S. are metaphorically 'in the corner of a workshop' in their domestic political system, but are expected to play the role of an Ivory tower internationally given the requirements of the diplomatic system for authorized and unified diplomatic positions.[26] As will be shown in the chapters 4, 5 and 6, differences between countries pertaining to the character of the state translate into variations in organizational arrangements of foreign affairs administration. Understanding these differences in institutional settings is of crucial importance when we analyze effects of IT-implementation and use at foreign ministries.

2.4 *Conclusion*

Diplomacy in its modern form emerged along with the emergence of sovereign states in Europe between the 15th and 18th Century. Throughout this process the right to send and receive ambassadors and thereby to create, receive and store authorized information pertaining to the foreign affairs of a particular state gradually became the exclusive prerogative of sovereigns. Principles that lay at the core of organizing interstate diplomacy—hierarchy, secrecy and one-way communication with the public—reflect the sovereigns' claim to sovereignty in conducting foreign affairs. Through continuous enactment of these principles, the role of diplomats as exclusive guardians of the system of state-to-state communication is maintained and thereby the power and authority of the state in the realm of foreign affairs (i.e. the 'informational prerogative') is sustained.

In an information-intensive environment brought about by easy availability of IT-tools, fast flows of information in various formats over the Internet and the ongoing media explosion, the amount of information that foreign ministries have access to on a privileged basis is radically shrinking. Non-state actors, private corporations and individual citizens empowered by the technology now often find out about international events simultaneously with or earlier than foreign ministries. What is

[26] The metaphorical imagery of the foreign ministry as 'an ivory tower' or as being 'in the corner of a workshop' was taken from Maybee, J. (1980): "Foreign Service Consolidation" *International Perspectives*, July/August 1980, p. 17, cf. Eayrs 1982:106.

more, foreign ministries are in an increasing manner dependent upon timely informational resources and expertise from non-diplomatic sources. The core of the IT-challenge to diplomacy hence lays in its undermining the basis of the state's 'informational prerogative' in foreign affairs. The question arises whether the technology-driven challenge might lead to radical transformation of the way diplomacy is organized or whether the structures, norms and procedures of diplomacy endure as they have for several centuries. As Nye (2002:62) argues, with the information revolution we may be moving towards a new "cyber-feudalism with overlapping communities and jurisdictions laying claims to multiple layers of citizens' identities and loyalties. In short, these transformations suggest the reversal of the modern centralized state that has dominated world politics for the past three and a half centuries". Given that the modern organizational basis of diplomacy has co-evolved with the wider institutional structure of the modern state order, the change eventually brought about by the implementation and use of IT at foreign ministries, might have various implications for the institutionalized mechanisms of interstate communications.

The next chapter discusses IT-revolution as a comprehensive challenge to the modern organizational basis of diplomacy and outlines an analytical framework for analyzing its change.

CHAPTER THREE

ANALYZING IT-EFFECTS IN FOREIGN MINISTRIES

3.1 *Introduction*

As the previous chapter illustrates, foreign ministries feature a number of institutionalized organizational specifics that differentiate them from other governmental agencies. While this is the case, they are susceptible to the same environmental changes related to the current rise in importance of information technology in various societal sectors. For foreign ministries, though, the current wave of innovations in information and communication technologies does not represent an entirely unprecedented development. Following the introduction of the telegraph in the 1850s, some radical reformers in the major European capitals kept announcing the demise of the expensive system of resident embassies and profound changes in the working methods of foreign ministries were anticipated. Although, changes did follow, they had the character of path dependent adaptations and no radical overhaul of the way diplomacy was organized and conduced had happened. A hundred years after the introduction of the telegraph, Garrett Mattingly (1955:17) could hence argue that numerous practices, procedures and structures of 15th Century diplomacy had survived the disintegration of the medieval order and could be recognized in the diplomatic practice of his time. Foreign ministries have proven highly adaptive to the changes in the organizational environment (Hocking 1999, 2007, Melissen 1999). In many respects, the ongoing information revolution comprises a profound set of challenges to the way foreign ministries are organized, raising questions about their organizational adaptive capacity. This chapter develops an analytical framework for studying this process.

In Section I, the chapter will, first, briefly address the way how foreign ministries have dealt with the communications revolution in the mid-19th Century related primarily to the introduction of the telegraph. This will be followed by a discussion of the broad based change dynamics that amount to what is currently often referred to as the information revolution. The discussion will first review some of the key contributions on the impacts of the information revolution on governments, which will be followed by a discussion of the changes in the operational environment of foreign ministries and the ITs that have been most commonly used in these agencies. This general discussion sets the stage for the development of the analytical framework that this study applies to analyze the effects of IT on the organizational basis of diplomacy.

In Section II, IT-supported organizational change is discussed in theoretical terms with a focus on the social embeddedness of IT. The core organizing principles of diplomacy identified in the previous chapter are then operationalized and three scenarios of IT-effects on the organizational basis of diplomacy are elaborated. Finally, the chapter addresses the choice of the studied cases and the research methodology applied. Conclusions follow.

3.2 *Foreign Ministries and the Information Revolution(s)*

From the perspective of foreign ministries, the current information revolution is not an entirely unprecedented kind of development. The industrial revolution between the 1830s and 1860s brought about the introduction of steamships, railways, and indeed of electric telegraph, which allowed messages to be transferred between the European capitals in a matter of hours.[1] As Jones (1983:116) argues, this fairly radical speeding up of communications "did not simply transform diplomatic techniques; it ultimately can be said to have fundamentally altered the character of diplomatic representation." The transformation could in simplified terms be presented as a gradual change of the role of the ambassador from an independent and influential figure with a great leeway to conduct foreign policy towards a "self-effacing, subordinate, and anonymous" official ready to execute the policy of the foreign minister (ibid.). In particular, the introduction of the telegraph was crucial in this development paralleled by increasing bureaucratization and centralization of foreign ministries and their networks of missions abroad.[2] In 1853 first telegraph links were established between

[1] In 1824, for a message from London to reach Vienna it would take between one and two weeks, while to Constantinople and St. Petersburg it would take between two and four weeks, depending on the weather conditions (Jones 1983:118).

[2] These organizational developments at foreign ministries are discussed in more detail in Chapter 2.

London, Paris and Berlin. The fact that instructions could now be sent to ambassadors on a daily basis even to more remote countries meant that centralized supervision and decision-making by the headquarters limited freedom of action of individual ambassadors. In 1861, some pragmatic reformers in Britain went even as far as to claim that the new technology (telegraph) had made the expensive embassy obsolescent (Hamilton and Langhorne 1996:132). Diplomats found it a challenge to adapt to the use of telegraph, mainly due to it seemingly undermining the long established procedures and routines of diplomatic practice. As Lord Hammond, the then British Permanent Under-Secretary of State, argued in a private letter in 1858:

> I confess I dislike the telegraph very much. In the first place, nothing is sufficiently explained by it. It tempts hasty decisions. It is an unsatisfactory record for it gives no reason. It is bad for the minister who leaves no justification of the ground of the decisions and it is inconvenient in a government like ours to have insufficient means of placing facts and arguments before parliament (cf. Jones 1983:123).

Besides this challenge to the bureaucratic procedures and record keeping, the telegraph was also challenging the independent role of the diplomat as a plenipotentiary negotiator. As one of the senior diplomats in the British foreign service had argued in 1902, this process had brought about "a telegraphic demoralisation of those who formerly had to act for themselves and are now content to be at the end of the wire" (cf. Anderson 1993:118). As Bátora and Neumann (2002) have shown and as the case studies in the following chapters will reveal, many a similar lament regarding e-mail communication could be heard from the diplomats in recent years following increasing electronic integration between headquarters and missions abroad. But coming back to the time of when the telegraph was introduced, the central argument of economy-minded reformers in the 1860s was that the telegraph enabling speedy communication of instructions to embassies would diminish the need for expensive senior ambassadors, who could instead be replaced by more junior diplomatic officers, who would simply relay information to the appropriate counterparts in the country of posting (Jones 1983:126). Yet, as a number of diplomatic incidents in the 19th and early 20th Century have shown, the speed of communications made the presence of highly trained and experienced diplomats more and not less important for successful negotiations (ibid.). On the other hand, the use of telegraph had become such a standardized part of diplomatic practice by the advent of World War I. that the Austro-Hungarian Empire found it appropriate to send a simple telegraph message to declare war on Serbia in July 1914 (Hamilton and Langhorne 1995:134).

Compared to the introduction of the telegraph, the current information revolution represents a far more comprehensive challenge to the established standards for organizing diplomacy as it is combined with a broader set of developments challenging the role of the sovereign state as the key actor in international relations. When characterizing these broad change dynamics, authors often refer to the 'information revolution', the 'post-industrial society' or the 'information society' (e.g. Bell 1974, Toffler 1980, Negroponte 1995).[3] Manuel Castells, who provided the academic community with one of the most comprehensive analyses of these complex processes (Castells 1996, 1997, 1998), argues that at present, we are experiencing a paradigm shift towards the *informational society*.[4] This term refers to a specific form of social organization, where information generation, processing, and transmission become the fundamental sources of productivity and power, due to newly emerged technological conditions (Castells 1996:21). The new technological conditions are brought about by massive penetration of information technology into virtually all spheres of our lives. In the current study, the use of the term *information technology (IT)* is identical with that of Groth (1997:27) and will hence broadly denote *all the technologies that depend on digital, electronic processing, including industrial automation, word processing, telecommunications and transmission of pictures and sound*.[5] Castells (1996:61–63) lists the following five features of what he sees as an emerging information technology paradigm:

[3] For a critical appraisal of the information society concept see Lyon (1988).

[4] Castells argues that this term better distinguishes the social and economic paradigm shift of the Information age than the more widespread term 'information society', which, as he suggests, can refer to most societies historically, as information plays an important role in any society.

[5] According to Preston (2001:25), the term information technology first emerged in the social science literature in the early 1980s, when personal computers became increasingly common in the workplace and in households. While the term *information and communication technology (ICT)* has been increasingly common since about the mid-1980s stressing the communicational elements of the new technological tools, I deliberately choose to stick to the more simple IT-term which, as might be clear from the above definition, encompasses communicational aspects. Some quotes used here, however, will contain the ICT acronym as it appears in the original texts quoted from.

1) There is an unprecedented *pervasiveness of IT effects* because information is an integral part of virtually all human activity;
2) IT supports a *networking logic*, which can be implemented in various kinds of organizations and processes;
3) IT provides *flexibility*, where not only processes can be reversed, but organizations and institutions can be 'reconfigured' by rearranging their components and adopting new elements;
4) There is increasing *convergence* of specific information technologies into a highly integrated system;
5) Information is the raw material of the new technologies and IT is a *technology to act on information*, which means ITs are processes to be developed and not just physical tools to be applied, and thus the human mind is a productive force, and not just a decisive element of the production system.

Information and its control have always been central elements in governmental organizing. In modern states, information pertaining to the government of a particular territory had been administered by bureaucracies with clearly defined jurisdictions, hierarchical line relations, functional specializations and formalized procedures ensuring impersonal machine-like execution of tasks (Weber 1978). A number of authors have suggested that these principles have been challenged by the rise of IT.

Frissen (1997:120–122), for instance, proposed that the IT-driven transformations have been unfolding in four broad tendencies. Firstly, *horizontalisation* involving increased IT-enabled horizontal network connections between individual units within governmental organizations. Secondly, *deterritorialisation* owing to the fact that with IT-supported provision of governmental services neither time nor space pose limitations, and governmental service provision can hence be organized "nearly independently of existing patterns and structures of public administration". The third tendency is *virtualisation*, which means that advanced IT applications are increasingly able to simulate physical realities, which further strengthens the tendency towards decoupling public administration and physical space (and/or physical infrastructure of public offices). And finally, Frissen also sees the IT-revolution as leading to *redesign*: public organizations having invested time and resources into setting up advanced IT-infrastructures are prompted to move ahead with deep organizational transformations so that the inherent potential of the technological applications in place can be capitalized upon.[6]

In another account of the ongoing IT-challenges to governmental organizing, Zuurmond (1996, 1998), suggested that bureaucracies are being replaced by what he calls *infocracies*, which denote administrative systems where the traditional control through bureaucratic structures is substituted by control through information infrastructures. Furthermore, increasing IT-connectivity, enabling integration of administrative processes across government, is seen as challenging established organizational boundaries of public agencies (Bekkers 1998). Environments can be colonized by IT-standards and procedures applied by a particular organization; organizations can be penetrated by technological standards from the environment; and new cross-organizational connections can be generated, leading to horizontal integration between organizations. This is said to have radical implications for legal and political accountability of public organizations and might lead to profound transformations of organizational structures of public administrations (ibid., Zouridis 1998).[7] While most of these challenges to governmental bureaucracies concern also the foreign ministries, these organizations also face a number of specific challenges related to the information revolution.

The new characteristics of the current environment relate firstly to *speed and simultaneity of global events*. A good example in this context is an event described by George F. Will in The Washington Post on October 10, 1996 (cited in the EFF Report 1998:25): After an overnight frost in Brazil, a local governmental official announced a substantial reduction in planned coffee production. The information instantly reached the Coffee, Sugar and Cocoa Exchange, where the prices of coffee immediately began rising. Then,

> traders of other... products did not understand what was happening but began bidding up their prices, causing the index of commodity prices to rise. This was registered on the computer screens... in almost two hundred Wall Street firms, who reported this shiver of inflation to their

[6] Arguing along the same lines in another contribution, Frissen (1998:39) went as far as to propose that in the light of these developments "departments will lose meaning".

[7] Obviously, the literature on the impacts of IT on public administrations reviewed here is only but a small fraction of the vast body of literature addressing the topic, and a wide array of other implications were proposed and observed when it comes to increased use of IT by governments. See for instance the volumes edited by Snellen and van de Donk (1998), Loader (1999), Heeks (1999), Kamarck and Nye (2002), and Slaata (2002).

> bond trading colleagues, who started a sell-off of bonds, which caused bond prices to fall, which caused bond yields to rise, which put upward pressure on interest rates, which caused stock prices to fall.

The time difference between the announcement in Brazil and the shock wave on Wall Street was less than ten minutes. In this kind of an international environment, many a diplomatic analysis has to be done in a network of several counterparts placed in different locations (having access to different sources of information, including each a different set of personal contacts) interacting as a virtual team. While advanced IT is obviously a sine qua non condition for any kind of global integration of diplomatic expertise, IT as such will not replace qualified personnel. Effective use of IT hence involves not only the technological hardware and communication systems, but also qualified people "who are able to take the lead and use information in innovative ways" (ibid., p. 23) This involves also the need for advanced training of the Foreign Service personnel in new management techniques including working in virtual teams.

A further characteristic of the current environment is the *proliferation of new foreign policy actors* (e.g. NGOs, IGOs, transnational corporations, media, global city regions, national banks and others), new foreign policy agenda (e.g. human rights, environmental protection, poverty issues, AIDS) and new threats to national security (global terrorist networks, hacker groups etc.). At a more general political level, as Frissen (1997:118–119) notes, societies in advanced welfare states have become 'smart', and it is no longer possible to steer them by direct hierarchical top-down governmental regulation. Instead, there is a trend towards process-oriented network-based consultative co-production of policies involving multiple actors creating public-private-NGO partnerships. Services and goods produced by these networks are often no longer clearly public or private; they are what Arquilla and Ronfeldt (1997:35) call collective. These development tendencies represent a complex challenge to the state-centric world order (Wolfish and Smith 2000, Nye 2002). Success of the International Campaign to Ban Landmines strongly supported by among others Canada and Norway shows that an extensive network cooperation of state actors with NGOs, pressure groups, activists and media can create positive outcomes in fairly short time horizons (Price 1998, Axworthy 2003, Sending and Neumann 2006, Hynek 2008).

Thirdly, in an information intensive environment, power no longer resides solely in governmental chanceries but increasingly in what could be called the global sphere of the mind, the *noosphere* (Ronfeldt and

Arquilla 1999).[8] Policy conducted in the noosphere does not entirely depend on traditional strategic resources such as military or economic power, but rather on the promotion of norms and values, the so called *soft power*,[9] which suggests a move from the classic *realpolitik* towards what Ronfeldt and Arquilla (1999) have called *noopolitik*. Noopolitik is defined as "an approach to statecraft, to be undertaken as much by nonstate as by state actors, that emphasizes the role of soft power in expressing ideas, values, norms, and ethics through all manner of media. [...] While realpolitik remains steadfastly imbued with notions of control, noopolitik is less about control than "decontrol"—perhaps deliberate, regulated control—so that state actors can better adapt to the emergence of independent nonstate actors and learn to work with them through new mechanisms for communication and coordination" (ibid.). It is obvious that in a situation where noopolitik is the emerging paradigm in world politics, there are incentives for organizational change in foreign ministries towards structures that would enable horizontal coordination of international activities throughout the government.

Today, the main task of foreign ministries remains the conduct of foreign policy. To do this effectively, effective use of IT is inevitable. This was recognized by the authors of the 1998 'Equipped for the Future'

---

[8] The term is based on the Greek word "noos" ('the mind'), and was revived by Ronfeldt and Arquilla (1999) taking inspiration from the philosophical work of Pierre Teilhard de Chardin (1989[1940]). Teilhard de Chardin saw the emergence of *noosphere* as a stage in the evolution of Earth (following the creation of the lithosphere, atmosphere, biosphere etc.) and defined it as a "global realm of the mind", "a thinking circuit" or a planetary "consciousness". Ronfeldt and Arquilla (ibid.) further suggest that "(t)he noosphere concept thus encompasses *cyberspace* and the *infosphere*. Cyberspace refers to "the global system of internetted computers, communications infrastructures, on-line conferencing entities, databases, and information utilities generally known as "the Net". This mostly means the Internet; but the term cyberspace may also be used to refer to the electronic environments and critical infrastructures of a corporation, military, government, or other organization." The term infosphere has a wider meaning and "encompasses [the cyberspace], plus information systems that may not be part of the Net. [...] this often includes broadcast, print and other media [...], as well as institutions such as libraries, parts of which are not electronic" (Ronfeldt and Arquilla 1999). Paul Virilio has a different understanding of the term infosphere. He sees it as "the sphere of information, which is going to impose itself on the geosphere" (see Der Derian 1996). Virilio's understanding of infosphere hence corresponds to Arquilla and Ronfeld's term of noosphere, and it will be the latter that I will use further.

[9] Nye holds that "soft power is the ability to achieve desired outcomes in international affairs through attraction rather than coercion. It works by convincing others to follow, or getting them to agree to, norms and institutions that produce the desired behavior." See Joseph S. Nye, Jr., *Bound to Lead: The Changing Nature of American Power*, Basic Books 1990.

(EFF Report). Even a decade after its publication, their analysis of the challenges in using IT experienced by the State Department continues to be representative of the challenges faced by foreign ministries world-wide. The report sees IT as a way to improve policy, not merely as a communication tool. When IT is used properly, it can provide effectiveness improvements in organizing information in the embassies, in the headquarters, and throughout the interlinked global network of the foreign ministry (Rana 2007). Embassy web sites can take over some of the major representation functions in a foreign country and provide well-prepared information to the visitors in the Cyberspace, when created with sensitivity to the local culture. This idea, obviously motivated also by a search for ways to cut costs in increasingly strained budgets of foreign ministries, was followed up as one of the key points in the 2006 speech by Secretary Rice on "Transformational Diplomacy" (Rice 2006).

Through electronic interconnectedness, interagency operability can be improved. For a diplomat, access to information is crucial and this is extensively provided through the use of IT. In addition, IT helps him/her in prioritizing information and making it broadly available at an almost instantaneous basis. Quick updating information and instantaneous countering misinformation about policies of the sending state to foreign journalists and decision-makers is another crucial asset (Grant 2005). Geographical distances are diminished since diplomats can attend meetings at headquarters or at international conferences electronically (Kurbalija 2007).

Concerns about secrecy and security of communication links remain a highly relevant issue and new kinds of threats emerge due to IT. In June 2007, for instance, alleged Chinese hackers were reported to have gained unauthorized access to the computer networks of the German Prime Minister's office including the private e-mail inbox of Chancellor Merkel, as well as to the networks of the German foreign ministry. This series of incidents involved also remote electronic infiltration of the UK Foreign Office and the Pentagon, where parts of the unclassified network had to be shut down for a week of repairs.[10] Cyberthreats to information security can also be perpetrated by single individuals. A

[10] See "Chinese See Dependence on Computers as Weakness" in *International Herald Tribune*, August 29, 2007; "China's Cyber Spies" in *Financial Times*, September 4, 2007; "China Hacked into Pentagon Defence System" in *Financial Times*, September 5, 2007.

21-year old Swedish computer security specialist demonstrated this in August 2007, when he set up his own node in a peer-to-peer network used by foreign ministries to exchange unencrypted information and hence collected thousands of partly sensitive e-mails and passwords from embassies of countries such as Russia and India.[11] Yet, although threats to information security proliferate, the traditional response of foreign ministries, i.e. complete "risk avoidance" by not being connected to the Internet is obviously not a viable solution, albeit perhaps one desired by many an information security expert. Instead, as the 1998 EFF report had found, foreign ministries are better advised to rely on 'risk management', because in an information intensive environment it may turn out to be riskier not having proper access to information than losing some particular piece of information.

Table 2 summarizes some of the most common uses of IT by foreign ministries.

Table 2: IT and its potential uses by foreign ministries (based on the EFF Report 1998:22)

Information technology and tools:	*Possible use by a foreign ministry:*
Common desktop environment	An advanced computer with a common suite of software and identical configuration of that software on each machine
Internal and external global networks	a) Secure communications worldwide (phone, fax, e-mail and video) b) Better communication between the headquarters and embassies and among embassies in the same region c) Direct communication between the headquarters and the foreign counterparts

[11] See "Security Researcher Intercepts Embassy Passwords" in *Computerworld*, September 10, 2007; "Tor Network Exposes Sensitive Information" in *The Inquirer*, September 10, 2007.

Table 2 (*cont.*)

Information technology and tools:	*Possible use by a foreign ministry:*
Databases	a) Retrieval, corporate memory, quick reference to previous work on selected topic b) Development of a global international affairs resource database c) Participation in the Geographical Information System (GIS)
Access to the Internet with web browsers and facilities to host web pages	a) Web pages for embassies b) Information service for citizens at home and abroad c) Public diplomacy on line d) Electronically provided information and links for commercial users e) Worldwide information searches
Automated message handling	Record-keeping, retrieval based on message title, subject, date, text search; ability to automatically separate unclassified from sensitive messages
Business process automation	Electronic payroll, bill paying, travel vouchers etc. at the headquarters and in the embassies
Collaborative tool software	a) Internet chat rooms to communicate globally and reduce number of time-consuming briefings and meetings locally b) Improved planning regionally and between the MFA and individual embassies c) Improved crisis management
Presentation software	Illustrated electronic memos
Video teleconferencing	a) Real-time collective analysis across global distances b) Long-distance interaction among the embassies and with the MFA c) Interaction with foreign counterparts
Digital photography	a) Rapid transmission of photographs b) Medical evaluation c) Personal identification d) Geographic markers e) Integration of photos with all messages or presentations
Commercial encryption	Tunneling secret information through unclassified channels

Obviously, the various uses of IT listed in the right hand column of the table above are only possible, if actual organizational procedures and routines are put in place enabling such uses. For instance, while videoconferencing software and hardware might be installed on computers in embassies and at headquarters of a given diplomatic service, 'real-time collective analysis across global distances' will not be performed unless the diplomats are IT-savvy enough to use the tools in effective ways and unless there are procedures that encourage such collaborative efforts. The same applies to implementation of 'collaborative tool software' which might lead to 'improved crisis management' only when appropriate procedures are put in place. As the EFF Report (p. 23) points out,

> [i]nformation technology is just that—a technology. Properly understood, information technology is not merely computers, software and communications links, but also the people who are able to take the lead and use information in innovative ways.

How then do we analyze the effects of IT on the organizational basis of diplomacy—the collectivity of foreign ministries? The next section develops an analytical framework to address this question. First, IT-supported organizational change is discussed in theoretical terms. Second, these theoretical ideas are combined with the conceptualizations of the three core organizing principles of diplomacy developed in Chapter 2 and analytical operationalizations of these principles are suggested. This is followed by the elaboration of three developmental scenarios of IT-effects on the organizational basis of diplomacy. Third, a methodology for studying IT-effects in foreign ministries is discussed.

3.3 *IT and Institutional Change Dynamics*

Most contributions addressing the impacts of technologies on organizations subscribe to the notion that IT is not a transformative force in and of itself, but should be seen as embedded in social contexts with which it co-evolves. In other words, IT and social settings are mutually constitutive (Barley 1986, 1990, Powell 1987, Bijker et al. 1987, Bijker and Law 1992, Edge [1987] 1995, Castells 1996, Preston 2001, Baldersheim 2004). As Groth (1997:449) points out,

> technologies can only be understood as part of a greater social context. [...] Social structures and technologies are therefore parts of the same

> continuum, all shaped by human action and developed, sustained, or obviated by the actions of innumerable individuals[12]

Using this general understanding of mutual embeddedness of IT and social structures as a point of departure, Bellamy and Taylor (1996) set the focus specifically on governmental structures and proposed the notion of *information polity*.[13] In their view, information and the media through which it is carried are central "to the definition, maintenance and reform of the institutional order which it reflects and represents" (ibid., p. 57). As they further elaborate, it is important to understand,

> how information media may both reflect and embed the rules of the game—the institutionalized routines—which govern how political processes occur. That is the distribution of informational resources is not simply to be understood as the outcome of the strategic behaviour of individuals. Rather, it is the outcome of systematic biases in institutions which are embedded not only in their formal, explicit procedural rules but also in their operating conventions and business systems. Thus, one of the major political issues associated with new ICTs is whether they will reinforce or challenge these biases, for example by modifying access to informational resources (ibid., pp. 57–58).

Hence, while IT-enabled institutional change dynamics is related to the technical aspects of available technologies, it is just as much related to non-technical aspects of the given information polity such as changes in established institutional rules, jurisdictions, norms, procedures and routines. As indicated above, institutions organize attention and set limits upon random change. Yet, if changes in the environment are sudden and extensive, institutional crises may unfold and changes may be radical. In the following two sections these two models of IT-supported institutional change dynamics are elaborated.

---

[12] This understanding is similar to the concept of *information system* (Orlikowski 1992, 1996; Avgerou 2001). As Avgerou (2001:46) points out, "what is generally called 'information system' in the jargon of practitioners as well as academics cannot be meaningfully restricted to computer or computer applications within an independently delineated social environment. Technical artefacts such as hardware, software, data in paper or electronic form, carry with them engineers with the conventions of their trade, industries that sell, install and support them, 'users' who understand their significance and interpret the way they should be put to action according to their circumstances and consultants who convert them from symbol manipulating machines to 'competitive advantage'".

[13] Bellamy and Taylor (1996) use the concept of polity in the Aristotelian sense as a stable, legitimized form of governance where political behavior is regularized by long-established norms, rules and structures.

3.3.1 *IT-enabled institutional transformation*

As indicated in some of the literature referred to in part 3.2, implementation and use of IT may lead to radical institutional transformation of governmental institutions, i.e. replacement of an established institutional order by a new one. This is a process that corresponds with what Olsen (1992) called *change of an institution.* Mechanisms related to IT driving such radical change may be found at the three levels of institutions described by Olsen (ibid.)—environments, structures and actors.

Firstly, standardized responses and routines ensure continuity and stability of organizations in a constantly changing environment. However, when changes in the environment are as rapid and radical as some of the technological changes in the recent decades have been, routine responses to such newly evolved conditions may lead to severe inefficiencies and a sense of *performance crisis* may emerge. In such situations, a problem-solving activity is initiated with the goal of constructing a definition of the new situation and of a new set of appropriate responses to that situation (March and Simon 1958:140). New IT-systems in public organizations are often put in place in particular organizational units for testing during a pilot period. These organizational units then develop appropriate procedures and performance programs which feature sets of routines and rules of how the new IT system will be used. This might involve new communication-patterns, involvement of new actors in information sharing, reshuffling of informational jurisdictions etc. Eventually, such a search process might lead to a lasting change of the existing repertoire of rules, routines and procedures underpinning a particular institutional setting.[14]

Secondly, institutionalized structures guide actions of individuals in organizations by decreasing complexity and indicating appropriate actions in given situations. IT implemented into an institutional setting might lead to ambiguity as to what is appropriate action.[15] IT might enable actions which challenge established ways of doing things and this might result in what Olsen (2004) had termed *institutional collisions*.[16]

[14] Search processes as such may be highly institutionalized in organizations (March and Simon ([1958] 1993:161), and development and implementation of new performance routines will therefore in such cases *not* result in radical change.

[15] For instance an e-mail system enabling instant communication irrespective of hierarchical line relations introduced in a traditional governmental bureaucracy might lead to such uncertainties.

[16] For views on the notion of *institutional collision* see also Thelen (1999), Clemens and Cook (1999), Orren and Skowronek (2004).

These are situations, where several logics of appropriateness might be evoked and actors are forced to (or as Friedland and Alford [1991] would argue, deliberately decide to) choose between competing institutional sets of criteria guiding action. Uses of IT in a particular way might contribute to a lasting shift in the application of notions of appropriateness, which then might lead to a lasting institutional change. Also, as Friedland and Alford (1991:255) argue, "[w]hen institutions are in conflict, people may mobilize to defend the symbols and practices of one institution from the implications of changes in others. Or they may attempt to export the symbols and practices of one institution in order to transform another." This applies within individual governmental agencies, but also to inter-agency relations. In the latter case, particular uses of IT in some parts of a government might challenge institutionalized practice and norms in other parts of the same government, and the organizations that are challenged then mobilize resources to create normative standards and government-wide rules that would ensure the maintenance of their institutional setting. Implementation and use of IT serves in such situations as a catalyst or trigger of institutional defense reflexes and counter-reactions, which might lead to introduction of new legal norms and standards.

Thirdly, the perceived transformative potential of IT has been inspiring governments world-wide to launch extensive reform-initiatives (Heeks 1999). In general, these approaches are based on the notion that an instrumental IT-savvy *change management* can lead to the introduction of new sets of IT-supported norms and rules. Ideas of business process reengineering (BPR) have played a primary role in this respect. Hammer and Champy (1993:31–32) refer to BPR as "tossing aside old systems and starting over" and define it more properly as "the fundamental rethinking and radical redesign of business processes to achieve dramatic improvements in critical contemporary measures of performance, such as cost, quality, service, and speed."[17] Reform initiatives introducing IT have been common at government-wide level, at the level of individual

[17] Given the specifics of the public sector, Andersen (1999) suggested that it could be counterproductive to simply transfer BPR ideas into public sector reforms. Instead he proposed the concept of *Public Sector Process Rebuilding (PUPREB)*, which better suits the needs of the public sector. The latter perspective alerts us to the fact that specific features of a given institutional setting may impose specific constraints upon processes of IT implementation. For the specifics of IT-reform initiatives in governments as opposed to the private sector see also other contributions in the volume edited by Heeks (1999).

public agencies, but also within public organizations. The latter involves processes where individual units within a public agency develop their own IT-supported working procedures and become IT-champions.

3.3.2 *Institutional resilience*

When an IT-tool is used by members of an organization, it comes into an internal environment of rules, norms, routines and organizational culture which influence how technology is used, for what purpose, and which of its aspects are used more and which less. Institutional and organizational arrangements mediate the meanings associated with objective technologies and their uses by organizations. The processes when technologies are made sense of and applied by organizational members can be called *technology enactment* (Fountain 2001:10–14, 88–90). As Fountain (ibid., p. 90) points out:

> The process of enacting technology refers to the tendency of some organizational actors to implement new IT in ways that reproduce, indeed strengthen, institutionalized sociocultural mechanisms even when such enactments do not use technology rationally or optimally. Organizational actors tend to enact technology to preserve ongoing social, or network, relationships and to maintain performance programs: the routines, scripts, frames, and patterns that constitute the typical organized set of responses within organizations.

What is crucial for understanding the enactment process is the *selectiveness* of attention paid to various aspects of technology. Facing a broad range of possible uses of a given IT, actors in an organization tend to choose those that match existing rules, routines and procedures. Technology is hence applied in ways that enable the maintenance of appropriate behavior related to the role of the given individual, of the organizational unit or indeed of the organization as a whole. In other words, only those aspects of IT are made use of that match the existing institutional setting. This does not mean that there is no change dynamics at play. Structures, procedures and routines are adapted to the new technological conditions, but the basic notions of what an organization is, what it does, what is its role and what behavior is expected of it remain unchallenged. Such path-dependent and gradual adaptation, where IT is used in ways that perpetuate the existing institutional setting corresponds to what Olsen (1992) had referred to as *change in an institution*.

Figure 2 summarizes the notions presented here about IT-enabled institutional change dynamics.

Figure 2: IT and institutional change dynamic

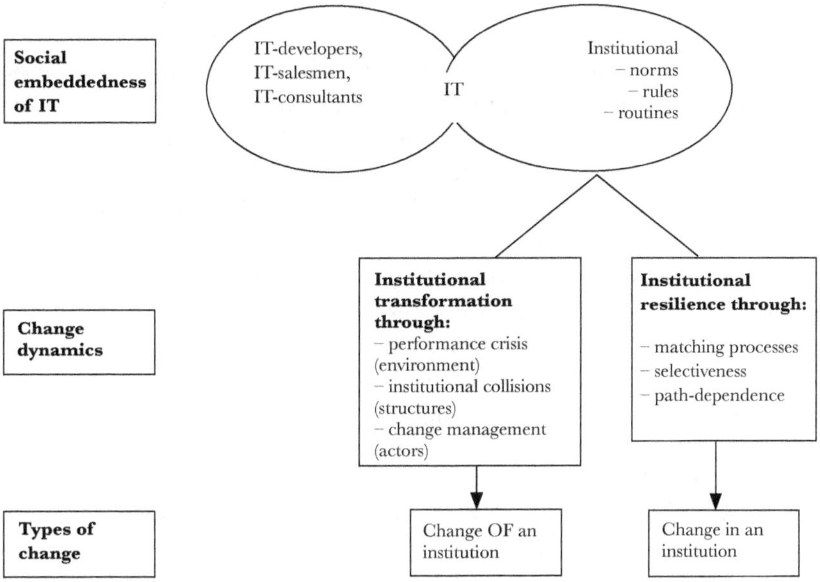

3.4 *Operationalizing the Core Organizing Principles of Diplomacy*

To analyze the effects of IT on the organizational basis of diplomacy, operationalizations of its core organizing principles need to be developed by identifying structures, procedures and practices of foreign ministries, which might be said to be organizational carriers and expressions of the principles. Secondly, the operationalizations need to be testable in relation to the effects of IT.

Reviewing the central analytical contributions by practitioners on the IT-related change dynamics in organizing diplomacy,[18] the recurrent themes include problems with the hierarchical structures and procedures in foreign ministries; their culture of secrecy; and indeed their perceived inability to communicate with the general public effectively.

The issues that one most often cites in relation to *hierarchy* is the bureaucratic character of working procedures at foreign ministries, which slows down the flow of information across the boundaries of

[18] Drawing here primarily upon the following reports: *From Quill Pen to Satellite* (Eldon 1994), *Equipped for the Future* (1998), *Reinventing Diplomacy in the Information Age* (1998), *The Electronic Foreign Service* (1999), *e-diplomacy: the FCO e-business strategy* (2001).

organizational units, across authority line levels and creates hindrances for including external actors in the foreign affairs community in information-sharing systems and shared databases. Secondly, the potential of IT for integrating organizational actors in the foreign service on a global scale are also cited, and foreign ministries' traditional clinging to the hierarchical relationship between headquarters and embassies is criticized. Hierarchy will hence in the context of the current study be operationalized at two levels. *A) Bureaucratic hierarchy* epitomized in the classical Weberian model of bureaucracy, where every bureau of the foreign ministry is constituted by a set of written files; there are clear divisions between official jurisdictional areas of offices; and there is an established system of super- and subordination of offices and hierarchical lines of responsibility. *B) Hierarchy between headquarters and missions abroad*, where the traditional pattern has been for the embassies to gather and report information and for the headquarters to perform analysis and issue instructions.

As pointed out in Chapter 2, *secrecy* has been a central norm in organizing foreign ministries. As IT applications become increasingly prominent in diplomatic information exchange and communication of foreign policy relevant information more generally, questions are raised in the literature about the proper application of information security standards by foreign ministries. More specifically, this concerns the issue whether the traditional notion of risk avoidance is useful in the electronic environment, and whether the application of the traditional 'need-to-know' principle remains a necessary and relevant. Hence, in the context of the current study, secrecy can be operationalized in two ways. *A) Approaches to electronic risks*, which involves the basic notions of what constitutes an information security risk in an electronic environment and methods regarding the protection of classified and sensitive information from being accessed by unauthorized parties. *B) Policies of internal information access*, which involve norms, rules, regulations and practices regulating access to information within a foreign ministry. The traditional logic has been that of policing access even to unclassified information.

Thirdly, as discussed in the previous chapter, foreign ministries' *communication with the public* has traditionally been following the model of one-way and *ex-post* provision of information on foreign policy decisions to the public. Involvement of the general public in *ex-ante* discussions of foreign policy priorities or particular proposals of initiatives has not been usual. Yet, as the above mentioned reports point out, interactive web-sites, chat-rooms and e-mail enable a far greater involvement of citizens in the making of foreign policy. To operationalize foreign ministries' public communication, the current study will be looking at two factors. *A) Direction of communication*, where it will be investigated whether communication keeps to the traditional one-way information provision mode or whether other patterns emerge. Here, one can think of a continuum of modes ranging from mere ex post provision of information to the public, through consultation of the public to actual engagement of the public in foreign policy processes (see Figure 3).[19]

Figure 3: Continuum of modes of diplomatic communication with the public

Information	→	**Consultation**	→	**Engagement**
(one-way press releases)		(the public encouraged to voice opinions on foreign policies)		(the public invited to review policy drafts and have their opinions considered before policies are adopted)

B) The second dimension will be *the level of centralization of public communication*, where the traditional model has been for foreign ministries to communicate with the media and the general public via a designated centralized unit, usually the press department, or indeed through a spokesperson or the minister.

Using the above developed operationalizations, an ideal typical traditional foreign ministry could be modelled in the following manner.

3.5 *Multiple Scenarios of IT-effects on the Organizational Basis of Diplomacy*

To study the effects of IT on the core organizing principles of diplomacy reflected in organizational arrangements of foreign ministries, the current study follows the methodological approach of multiple competing scenarios suggested by Robey and Boudreau (1999). As their review of

[19] I am grateful to Evan H. Potter for pointing this out to me. See also Hay (2000) for a discussion of the differences between consultation and engagement.

Table 3: Operationalizations of the core organizing principles of diplomacy in an idealtipical traditional foreign ministry

Organizing principle of diplomacy	*Operationalization in a MFA*
Hierarchy	– Bureaucratic hierarchy – Hierarchy between HQ and missions abroad
Secrecy	– Risk avoidance approach to electronic risks – Information policing as the main policy of internal information access
One-way public communication	– One-way direction of communication; ex-post communication – Highly centralized public communication (e.g. via Press Department)

the literature on IT-driven change dynamics in organizations suggests, contradictory findings about IT-effects are reported across studies of IT-implementation in similar kinds of organizations using similar kinds of technologies, and indeed also within particular individual organizations. They hence argue in favour of developing multiple hypotheses or propositions about IT-effects. Referring to the work of Chamberlin ([1890] 1965)[20] they point out that "entertaining multiple, competing hypotheses may protect researchers from too strong an affection for favourite theories" (Robey and Boudreau 1999:179). What is hence called for in the context of the current study is the development of a set of scenarios representing various possible patterns of IT-effects on the core organizing principles of diplomacy. Each of the scenarios would feature assumptions about the effects of IT on hierarchy, secrecy and one-way public communication as operationalized above (see Table 1 below for a summary of the scenarios). In Yin's (1994:106–110) terms, each of the scenarios will represent a particular change pattern or a *grouping of types of empirical variation.* These patterns will serve as a set of analytical devices helping us to categorize and organize empirical

[20] Chamberlin, T.C., 1890, "The method of multiple working hypotheses": *Science, Vol.* 15, pp. 92–96; reprinted 1965, Vol. 148, pp. 754–759.

evidence. Findings on the effects of IT at a particular foreign ministry will then be compared with the scenarios to identify the most plausible change pattern (this is conducive with the technique of *pattern matching* [ibid.] discussed below).

Using the above discussion of IT-enabled institutional change as a point of departure, effects of IT on the core organizing principles of diplomacy can either lead to radical transformation of diplomacy; adaptation of IT to the established institutional setting of diplomacy; or indeed a path-dependent change of diplomacy. These developmental options form the basis of the following three scenarios.

Scenario 1: Transformation

Effects of IT bring about a profound transformation of the organizational basis of diplomacy

At the level of individual foreign ministries, this scenario assumes that increased use of IT leads to a situation where bureaucratic line relations are undermined as official jurisdictional areas and office hierarchy is transgressed by IT-supported issue-oriented projects which become the primary organizing principle. Task-forces and virtual teams featuring members from throughout the foreign service (irrespective of geographic location) and external experts replace formal organizational units in the foreign ministry. This in turn leads to replacement of hierarchy as the main organizing principle by horizontal network structures.

Effects of IT also have implications for the hierarchy between headquarters and embassies: through virtual project cooperation, external actors become directly involved in the operations 'inside' the foreign ministry. Political responsibility for the various network projects then does not necessarily reside in the foreign ministry, but migrates to other governmental bodies which head the respective projects. Embassy personnel and headquarters staff participating in such projects then may be simply participants in projects politically directed by other governmental departments; hierarchical relations between foreign ministry headquarters and embassies are hence weakened. Through IT-supported horizontal network integration with other governmental bodies, NGOs, academia and other actors, diplomacy as an activity is no longer the exclusive prerogative of the foreign ministry, but becomes embedded in the mechanisms that bind other actors to the foreign ministry in shared efforts to work on foreign policy goals. Norms of diplomatic appropriateness defining who is a diplomat, what are appropriate diplomatic situations and what is appropriate diplomatic conduct are hence transformed to include a wider array of actors.

Information security and secrecy in such a system of shifting network cooperation with external actors is transformed in terms of what is being protected and how it is done. In network projects it is not of primary importance to protect information at the nodes, but rather the 'embedded knowledge' present at the nodes of the network as a whole.[21] A foreign ministry involved in several network projects featuring external actors will then be part of several of such centers of gravity on which information security is based. While secrecy traditionally resided in the foreign ministry and it therefore had one center of gravity for information security, responsibility for secrecy now moves outside the foreign ministry to wherever a network project features sensitive or classified information. At the foreign ministry this would involve a situation, where some projects would be protected, while others would be kept open, allowing broad access to actors throughout the foreign service and also to external actors. The share of responsibility for secrecy hence both shrinks internally at the foreign ministry and extends beyond its boundaries to other actors. The approach to electronic risks emerging in this manner could hence be labeled *risk export*.

Effective participation of external actors in network projects depends on having quick access to updated information. The need-to-know principle is substituted by open access to a pool of unclassified documents. It is considered central for ensuring connectiveness and consistency[22] of the network projects for all participating actors to have full access to unclassified information. Hence, the approach to internal information access is conducive with the notion of *information providing* (as alluded to in the EFF Report, p. 23). This involves a set of practices enabling broad access to electronic documents and files available throughout the foreign service with the goal of providing the broadest possible information resources for diplomatic analysis and decision-making.

---

[21] As Fast (1996) points out "corporate knowledge embedded in teams—like NASA's team that put man on the moon—is knowledge that none of the team members knows alone. Embedded knowledge is hard if not impossible to steal. Thus our experience and social networks that develop and use information technologies are precious commodities. We can identify them as our strategic center of gravity in the information age environment."

[22] For these terms see Castells (1996:171).

In the context of network cooperation with a broad array of foreign policy actors, IT-enabled communication with the public is embedded within the multi-actor diplomatic projects and not necessarily solely within the foreign ministry. In this situation, a request for information from the public directed at the foreign ministry or at one of its units may be redirected to the relevant cooperating agency or several agencies for feedback. Communication with the public is therefore *multi-channel.* This includes shared issue-related web-sites of the network partnerships, but also web-sites of partner organizations.

When it comes to the mode of communication with the public, the fact that external actors are directly involved in the projects with the foreign ministry or its parts, they also participate on reviewing drafts and participate in the formation of policy. Furthermore, the foreign ministry encourages the citizens at large to review policy drafts, come up with feedback and have their suggestions considered in the formation of foreign policy goals and priorities *ex ante.*

Overall, the effects of IT on diplomacy may include IT-driven changes from hierarchical arrangements towards horizontal networks; from secrecy towards openness and risk export; and from one-way communication towards multi-channel public communication. Such changes would amount to the formation of a *new information polity*[23] of the respective foreign ministry featuring new formal rules and operating conventions. At the level of individual foreign ministries such development might inspire reform at other foreign ministries through a wave of mimicry. This might have implications *at the level of diplomacy as an institutional system.* An alternative field of IT-supported cyber-diplomacy might eventually be structured, featuring new structures, actors, norms and procedures for organizing diplomacy. New notions would emerge regarding what a foreign ministry is and about what its role should be; about who and how is entitled to conduct diplomacy; and indeed about what constitutes diplomacy. In short, an entirely new pact of diplomacy as an institution with the respective national societies and with the system of interstate relations would be struck.[24] In practice, such a development leading to a convergence of foreign ministries around a new set of IT-supported norms, rules and principles could well be termed a *radical change of the organizational basis of diplomacy.*

---

[23] The concept of *information polity* is discussed in more detail in part 3.3.

[24] For the notion of a *pact* an institution has with the society and with other institutions see Olsen (2005:6–7).

Scenario 2: Status-Quo

Effects of IT do not Bring About any Profound Change of the Organizational Basis of Diplomacy

At the level of individual foreign ministries, this scenario presupposes that the process of IT-enabled organizational change is severely limited, incremental and path-dependent. What is more, there is both active and passive resistance to change. The overall pattern is that IT is used only to supplement paper-based practices. *Hierarchy* is maintained. This applies, firstly, to the bureaucratic line relations, where procedures, rules and routines correspond to the classical *Weberian model*. File management keeps paper-based standards, which limit the role of technology. Boundaries between official jurisdictional areas in the organization are carefully observed and maintained. Office hierarchy with a clearly established system of super- and subordination is strictly obeyed. Electronic communication does not happen in ways inconsistent with information flows on paper—whether this means not transgressing official jurisdictional areas between organizational units or not "jumping" across levels of hierarchy. Hierarchy is also maintained in relations between headquarters and missions abroad. Traditional relationships are maintained: embassies gather and report information, and headquarters perform analysis and issue instructions.

When it comes to *secrecy*, the approach to electronic risks remains *risk avoidance*. It implies that ensuring secrecy has an absolute priority in the foreign ministry, and risks of losing information to unauthorized parties are to be avoided wherever possible, even if this means that the foreign ministry's electronic systems will not be connected to the Internet and other external systems. This mindset is pervasive also in the approach to internal information access, which follows the principle of *information policing*. This implies that that the access of foreign service officers to (both classified and unclassified) documents is limited, controlled and policed. This involves a strict application of the *need-to-know* principle[25] regarding access-rights of foreign ministry staff.

---

[25] The need-to-know principle implies that an officer has access only to those files that directly concern his/her daily work. Most usually these are documents produced and managed by the specific organizational unit that the officer is assigned to.

Electronic *communication with the public* is consistent with the traditional notion that the communicating entity is the foreign ministry as a unified whole, be it through the foreign minister, the senior management or through a press secretary. Direction of communication is *one-way*—from the foreign ministry to the public. Documents displayed on web-sites communicate information to the public, but do not provide the possibility for feedback and interactive communication. The mode of communication with the public remains the traditional *press-release-mode* involving *ex-post* information about foreign policy decisions.

At the level of diplomacy as an institutional system, this scenario presupposes that the established institutional setting of diplomacy severely limits uses of IT to those aspects that do not challenge established notions of what diplomacy is and what role it plays. Institutional change that occurs is "likely to reflect local adaptation to local experience and thus be both relatively myopic and meandering, rather than optimizing, as well as 'inefficient', in the sense of not reaching a uniquely optimal arrangement (March 1981, cf. March and Olsen 2005:8). IT-driven changes at foreign ministries will be mediated by the local institutional settings featuring specific organizational histories, structures, actors and environments. This in turn will mean that differences between these organizations in terms of organizational structures and practices will continue to define them. At the same time, the institutional pact between diplomacy and respectively the national societies and the global society of states would remain unchanged. Using the analytical terms suggested above, this would in general amount to what might be termed *marginal change in the organizational basis diplomacy*, which would entail that existing relations of power and authority would be maintained.

Scenario 3: Renewal

Effects of IT lead to a path-dependent renewal of the organizational basis of diplomacy

At the level of individual foreign ministries, IT-enactment leads to re-assertion of the basic notions of what a foreign ministry is and what it does in the electronic environment using IT-tools. *Hierarchy* is revitalized. In terms of bureaucratic line relations, this involves the development of systems and applications for electronic management of documents and information, which are consistent with the professional diplomatic norms, rules and procedures, as well as the norms, rules and practices of bureaucracy. Most documents are exchanged and stored in electronic form. Official jurisdictional areas and office hierarchy are maintained, but they are often transgressed in cooperative efforts between officers from different units in the organization. IT-supported projects and task-forces complement standard line relations. Official files are stored in agenda-specific electronic databases across organizational units.

In terms of hierarchy between headquarters and embassies, there is integration of informational resources so that the foreign service works more like a *single global on-line organization*. Although embassies continue to be administratively placed under geographic bureaus in the headquarters, workload is shared and integrated. While embassies still gather and report information, they also perform analysis. As analytical tasks are moved out to embassies, they often task officers at headquarters as needed, and headquarters also performs information gathering on the ground in the home capital (for instance scanning the domestic political landscape for the viability of particular foreign policy initiatives in relation to particular countries that embassies might be working on) which they then report to embassies.

Secrecy remains a central norm in the foreign ministry's operations and handling of sensitive information. Yet there is an understanding that that although 100% security of electronically communicated and stored information cannot be guaranteed, there are greater risks stemming from not being connected to the Internet and not having updated information, than from the possibility that some individual pieces of information might be lost. The risk of losing a particular piece of information to unauthorized parties is outweighed by the benefits of having constant and broad access to timely and updated information. In terms of the approach to electronic risks, risk avoidance is replaced by *risk management* (as defined in the Equipped for the Future Report, p. 23).

Access to both classified and unclassified information is regulated by the electronic version of the need-to-know principle, but informational resources are organized in a fashion where information about information is generated. This means that there is a difference between access to information *about* documents, which is provided on a broad and open basis to employees of the foreign ministry, and actual access *to* documents, which is limited based on the need-to-know principle and other kinds of authorizations. External actors are not included in information-sharing systems. The approach to internal information access hence could be labelled *guarded openness*.

Communication with the public in cyberspace provides the possibility for feedback and interaction between the diplomats on the one hand and citizens, media and other involved actors on the other hand. While the communicating entity remains the foreign ministry, its communication efforts may be complemented by those of its organizational units or of individual officers. The direction of communication is *two-way*. The public is encouraged to provide feedback and ideas on foreign policy related issues, so that the mode of communication is *consultation*.

In sum, while this scenario assumes that IT-effects will lead to changes in practices, routines and structures at individual foreign ministries, the changes are not as profound as to challenge the basic notions of what a foreign ministry is and what is its role. Hence, at the level of diplomacy as an institutional system, IT-effects hence do not lead to a replacement of the existing set of norms, rules, routines and procedures of diplomacy, but to a reinvigoration of the existing norm-set using the potential of IT. This may be referred to as *extensive path-dependent change in the organizational basis of diplomacy* involving maintenance of existing relations of power and authority, but on renewed IT-supported fundaments—a renewed pact of diplomacy as an institution with individual societies and the global society of states.

Table 4 below summarizes the three scenarios of IT-effects on the organizational basis of diplomacy.

Following this outline of scenarios of potential IT-effects on the organizational basis of diplomacy, the next part outlines research methodology for studying the change dynamics in foreign ministries.

3.6 *Methodology*

To investigate effects of IT in foreign ministries, in-depth analyses of these processes within the organizations are called for. Such in-depth insights are necessary to produce meaningful assessments of the magnitude and the direction of change of the deeply entrenched institutional norms, rules, practices, routines at foreign ministries that diplomacy as an institution is maintained by. There are several reasons why a *case study method* is a suitable methodological approach for such an analysis. As observed by Yin (1989:23) a case study is an empirical inquiry that:

Table 4: Scenarios of IT-effects on the organizational basis of diplomacy

Scenario *Feature of diplomacy*	TRANSFORMATION	STATUS QUO	RENEWAL
ORGANIZATIONAL STRUCTURE	Network projects; HQ and embassies become part of shifting foreign policy networks	Weberian hierarchy; Hierarchy between HQ and embassies	Hierarchy complemented by flexible structures; Integration between HQ and embassies
SECRECY	risk export & information providing	risk avoidance & information policing	risk management & guarded openness
PUBLIC COMMUNICATION	project networks as communicators; multi-channel; engagement mode	MFA as communicator; one-way communication; press-release mode	MFA and its units/officers as communicators; two-way communication; consultation mode

- investigates a contemporary phenomenon within its real-life context; when
- the boundaries between phenomenon and context are not clearly evident; and in which
- multiple sources of evidence are used.

The effects of IT on the core organizing principles of diplomacy, are examined in the real-life context, namely in foreign ministries. As IT implementation and use represent integral parts of a broader set of organizational dynamics in foreign affairs establishments, one can say that the boundaries between the phenomenon and the context are not clearly evident. Multiple sources of evidence are used, including internal and public documents, and interviews with officers of the respective foreign ministry. Moreover, every foreign ministry is studied in its specific time and place setting, where data collections essentially focus on gathering empirical evidence related to IT-effects on the constitutive features of diplomacy.

To test the implications associated with the three scenarios, we need to use empirical evidence on effects of IT in foreign ministries with different institutional settings and different levels of IT-implementation. The goal is to test whether the effects of IT lead to convergence around a new set of IT-enabled diplomatic norms, rules and structures as envisioned in the transformation scenario; or indeed also whether the respective institutional settings prove resilient and differences in organizational structures between foreign ministries remain in place, as assumed in the status-quo scenario.

3.6.1 *Choice of cases*

As mentioned in the introductory chapter, the foreign ministries of Canada, Norway and Slovakia were chosen for investigation, due to their different institutional settings and differences in IT implementation and use to test the magnitude and direction of change.

Canada is a federal state which only gradually acquired sovereignty from its former position as a dominion of the British empire (see Chapter 4). Due to the specifics of the Canadian statehood and the national-administrative system featuring a plethora of internationally active actors, DFAIT has always had to struggle with other governmental departments, provinces, businesses and NGOs for its role as the coordinator of the country's foreign affairs (Stairs 1973, 2001; Nossal 1993, 1997, 2001; Cooper 1999, 2004; Michaud 2001, Axworthy 2003). Early adoption, implementation and innovative use of IT had been improving DFAIT's position in the continuous struggle for relevance as the coordinator of foreign affairs in the shifting networks of internationally active Canadian actors (Smith 1997, Smith and Sutherland 2002, Copeland 2005). Due to these prerequisites, DFAIT had been a *vanguard* world-wide in terms of applying IT in support of its operations (Potter 2001, 2002). It was chosen here to represent a case where the effects of IT on the core organizing principles of diplomacy were expected to be the most profound. This case was hence expected to play the role of a benchmark of the magnitude of change.[26]

[26] This is conducive with the view of Cooper (2004:253), who points out that due to its innovative ways of including civil society actors in the formulation and conduct of foreign policy, Canada serves as "a model of how foreign policy should be restructured [...and] if there is any momentum left at all for further innovation, the manner by which Canadian diplomacy meets the test will send us important signal about what things are happening—an how and where—on the front lines of global governance."

The N-MFA, as elaborated on in Chapter 5, is a century old foreign ministry of a unitary state with a continental tradition of sovereignty (see Huntington's ideas on the various ways sovereignty was institutionalized in various states in the previous chapter). It has been characterized by strong adherence to hierarchical line relations and bureaucratic procedures, a culture of secrecy, and reluctance to engage in broad-based public discussions of foreign policy priorities. Organizational reform initiatives have traditionally met with resistance and exploitation of established procedures and routines (Colding 1978, East 1981, Christensen 1996, Neumann and Leira 2005). IT-initiatives were no exception (Bátora and Neumann 2002). According to its self-assessment produced in 1999, N-MFA has been a serious *laggard* compared with foreign ministries of comparable countries in implementing and using IT in its daily operations.[27] The case of N-MFA was hence chosen in this study as a foreign ministry, where IT-effects on the core organizing principles of diplomacy were expected to be the least profound.

The third case chosen for investigation here was the *MFASR* representing a *new foreign ministry* of a new state founded in 1993. Although it was built around a small core of professional diplomats with experience from the Czechoslovak foreign ministry, the MFASR did not feature a well-established set of institutionalized routines and procedures that could be challenged by IT (Bátora 2000, 2003). Its structures, procedures, rules, norms and guidelines were being established in what might be termed an information age environment. The view of organizations as open systems dependent upon resources from and structured in interaction with their environment (see for instance Stinchcombe 1965, Hannan and Freeman 1977, 1989, Pfeffer and Salancik 1978, Hannan and Carol 1995, Baum 1996) would suggest that the MFASR was likely to be profoundly permeated by IT and by innovative ways of IT-supported organizing. On the other hand, additional insights from the new institutionalist literature exemplified by the work of Meyer and Rowan ([1977] 1991) and DiMaggio and Powell ([1983] 1991) would suggest that in the quest for legitimacy as a particular kind of organization (i.e. a foreign ministry), the MFASR will most likely adopt the prevailing models of organizing in the field of foreign ministries and will not get involved in much experimentation with innovative designs on its own. The case of MFASR was hence chosen for the current study to represent

[27] See the strategic report *The Electronic Foreign Service*, Oslo: MFA, June 25, 1999 (available online at http://www.regjeringen.no/nb/dep/ud/dok/rapporter_ploner/rapporter/2000/The_Electronic_Foreign_Service.html?id=420085

a foreign ministry where one expected to find either an entirely new set of IT-supported principles for organizing diplomacy, or a traditional set of structures and procedures for organizing diplomacy.

In sum, the choice of cases was informed by the expectations to find the most profound changes at DFAIT, the least profound changes at N-MFA, and the MFASR was expected to be between either one of these positions. Despite these differences in IT-implementation and use related to the particulars of the respective foreign ministry's institutional setting, the results of the case studies are by no means predetermined. Effects of IT, as pointed out in part 3.5 above, are mediated by the institutional setting of diplomacy, of which all the three organizations investigated are carriers. What we are concerned with here is whether IT-effects might lead to a profound transformation of core organizing principles of diplomacy.

3.6.2 *The case-study analysis*

The institutional differences between the cases provide a good point of departure for studying a potential transformation of diplomacy, because such a development would imply introduction of a new set of IT-supported structures and procedures at foreign ministries irrespective of previous differences in their institutional backgrounds. On the contrary, status-quo would imply local path-dependent adaptations and resistance to change, which would entail continued differences in implementation and use of IT.

Coming back to the above developed notions of change *in* and *of* diplomacy and to the theoretical approaches conceptualizing the interplay between IT and institutionalized settings, the status-quo and renewal scenarios represent change *in* diplomacy, while the transformation scenario represents change *of* diplomacy. While the status-quo scenario is consistent with the archetype features of the foreign ministry as outlined in Table 1 in Chapter 2, the two other scenarios represent clear logical breaks with that model. Each of the scenarios hence represents a particular *change pattern* in the sense of Yin (1994:106). Empirical data collected at the three foreign ministries in focus will be matched to the matrix of three change patterns (scenarios) in a process that Yin (ibid.) has called *pattern matching*.[28] For a graphical outline of the analytical process see Figure 3 below.

---

[28] A pattern-matching logic, as Yin (1994:106) elaborates, "compares an empirically based pattern with a predicted one (or several alternative predictions)." Essentially, one

Figure 4: The model for analyzing effects of IT at an individual foreign ministry

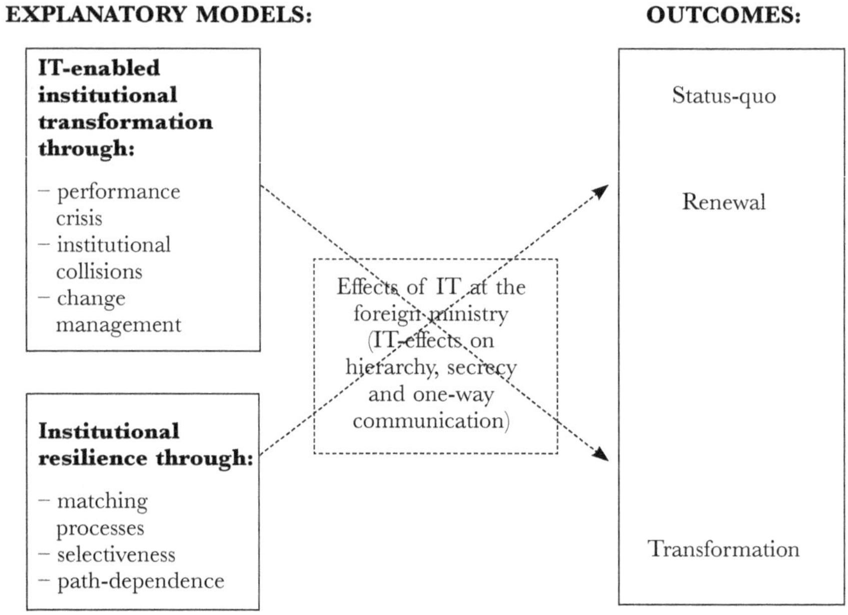

3.6.3 *Data material*

As discussed in Chapter 2, change *of* the organizational basis of diplomacy involves the introduction of a new role for the foreign ministry and a new logic of appropriateness guiding actions of diplomats including a new set of IT-supported organizational arrangements supporting the diplomats' operations. The goal of the data collections in the three foreign ministries was to find evidence of whether such a change is tak-

collects empirical material at foreign ministries by looking at the factors underlying the constitutive features of diplomacy reflected in organizational arrangements at foreign ministries (e.g. approach to electronic risks and approach to internal information access in relation to secrecy) and then match empirical findings to the three patterns of change anticipated in the scenarios. The boundaries between patterns are made clear by developing a fairly straightforward set of indicators, that determine what data collected in relation to the three non-equivalent variables would clearly align them with one of the three patterns. This is necessary, because as Yin (ibid. p. 34) suggests, "without any prior specification of the significant, operational events that constitute 'change', a reader cannot tell whether the recorded changes in a case study genuinely reflect critical events [...], or whether they happen to be based on an investigator's impressions only." In general, clear and comprehensive descriptions of the various patterns of change were hence needed.

ing place and/or to assess whether the magnitude of change is smaller. Although interviews with officers were conducted in all three foreign ministries, they were *not* directed at collecting information about the foreign ministry officials' perceptions of their role and identity. Such assessments were deemed as being to too personal and based on individual experience, and therefore hardly providing objective evidence of change tendencies. Instead, the focus of the data collections was directed at finding evidence of more objective indicators of IT-related organizational changes at the respective foreign ministries by looking at changes of structures, procedures and routines recorded in official documents.

The data that are used as a basis for the analysis consist hence of public and internal reports, memos and other documents from the three ministries. Some of these documents I copied in the respective ministerial library, some could be downloaded from the respective web-site, and still others I received from informants from within the respective organization. Information from the documents was complemented by *in situ* observation of internal IT-enabled practices and by a series of interviews with officers at each of the three ministries (see Appendix 2, 3 and 4 for lists of interviews featuring names and dates).[29]

3.7 *Conclusion*

Following up on the conceptualization of diplomacy as an institution developed in Chapter 2, this chapter has developed an analytical framework for assessing effects of IT on the organizational basis of diplomacy. It first addressed introduction of the telegraph in the mid-19th Century as a as the first communications technology challenge to the conduct and organization of diplomacy in modern times. This was followed by a discussion of current information technologies and of the challenges

[29] I used a semi-structured interview-guide featuring a set of open-ended questions focusing on the effects of IT on the practices, procedures and routines associated with the three core organizing principles of diplomacy—hierarchy, secrecy and one-way communication with the public. The interviews served hence the purpose of enhancing the information acquired through the study of official documents. I used the technique of *triangulation* (Yin 1994:91–93) to analyse the data collected through the study of reports and ministerial documents, through interviews and through in situ observations. In this way, I was able to use multiple sources of evidence in a corroboratory manner to assess the same phenomenon.

their rise represents for public administrations in general and for foreign ministries specifically. Mechanisms of IT-supported organizational change dynamics were then discussed in theoretical terms using insights from organizational theory and institutional analysis. Implementation and uses of IT in institutionalized organizations were discussed as leading to two types of institutional change. On the one hand, institutional transformation, which might occur if IT implementation is coupled with a sense of performance crisis, with institutional collisions or with change management. On the other hand, it was pointed out that established sets of norms, rules, routines and procedures might mediate and limit uses of IT and the result might hence be maintenance and persistence of established institutional structures.

Based on the discussion of the key characteristics of the organizational basis of diplomacy in Chapter 2, operationalizations of hierarchy, secrecy and one-way public communication were developed. Using the theoretical discussion of IT-enabled institutional change as a point of departure, three scenarios were then drawn up—transformation, status-quo and renewal—outlining possible effects of IT use and implementation in foreign ministries. The scenarios were organized around potential effects of IT on the three core organizing principles of diplomacy—hierarchy, secrecy and one-way public communication.

Three foreign ministries were chosen as cases for the investigation. The choice of cases was related to their different institutional backgrounds which had implications for variations in IT-implementation and use. DFAIT could be considered a vanguard in using IT, N-MFA a laggard in using IT, and MFA SR was a new organization. These differences were central in relation to the overall goal of this study which is to find what the effects of IT are on diplomacy as an institution. More specifically, the goal is to find whether effects of IT indeed lead to an IT-revolution in diplomatic affairs or whether the institutional setting of diplomacy is resilient and accommodates the IT challenge.

Before moving on to the presentation of the empirical findings, a final methodological remark is in order. As fieldworks at the three foreign ministries were conducted in specific time-periods, the data collected provides us with mere snap-shots into the respective organization at a particular point in time. It is hence a challenge to discern IT-related effects unambiguously as opposed to developments brought about by other institutional factors. Therefore, in the presentation of each of the three cases that follow, I supplement the specific study of the effects of IT on the three core organizing principles of diplomacy with a broader institutional depiction of the respective foreign ministry. This concerns primarily institutional characteristics related to *organizational structure, personnel and organizational culture*, and not least the *domestic environment* in the three countries.

Formal organizational structure and the professional backgrounds and socialization of personnel are aspects conditioning the execution of *hierarchical authority* within an organization (Egeberg et al. 1989, Christensen and Egeberg 1989, Egeberg 1999, Egeberg and Sætren 1999). Foreign ministries are no exception (Christensen 1996; see also the contributions in the volumes edited by Steiner [1982] and by Hocking [1999]).

Organizational structure and professional backgrounds of personnel also have implications upon the application of *secrecy* in a foreign ministry. Organizational structure channels formal information flows and in strictly hierarchical bureaucracies, secrecy is more easily maintained than in flat organizational structures with numerous participants and broad information sharing. Professional backgrounds and socialization of foreign ministry personnel also have implications for the extent to which a foreign ministry is characterized by what came to be termed the "culture of secrecy".[30] In general, shared educational background and/or specialized socialization techniques increase a shared identity among diplomatic personnel of a given foreign ministry and facilitate the drawing of boundaries between insiders and outsiders (Neumann 1998). On the contrary, greater heterogeneity in professional backgrounds and lack of socialization procedures will likely have opposite effects.

Finally, public agencies' practices of communication with the public vary in relation to their role and the role their agenda plays in the domestic political environment (Allern and Lorentzen 2002). A foreign ministry's role in the domestic political environment, and its relations with the domestic actors, has implications for the development of patterns of the foreign ministry's communication with the public (Smith and Sutherland 2002, Bátora 2006a).

Following this outline of the theoretical concepts and methodological approaches, we can now turn to the case studies. The first case to be addressed is DFAIT.

[30] This term was used in the 1998 *Equipped for the Future* report to describe the culture of the U.S. State Department.

CHAPTER FOUR

CANADA'S DEPARTMENT OF FOREIGN AFFAIRS AND INTERNATIONAL TRADE

We don't want the bodies to come into the office. We want the office to be in the bodies.[1]

4.1 *Introduction*

According to a study by Accenture, 2004 was the fourth consecutive year that the Canadian e-government program continued to set the standard for the rest of the world.[2] It is my contention that among foreign ministries worldwide, there was at the time of writing hardly a match to the Department of Foreign Affairs and International Trade (DFAIT) when it comes to using modern e-government solutions in support of its operations, although some serious contenders are emerging. From the early 1970s onwards, the organization has gone through a series of reorganization processes along with and largely due to the general transformation trends in the Canadian society. First, trade officials were included in order to regain legitimacy in the society at large and influence in elite circles (bringing more managerial techniques and business-like culture into the organization), then frameworks for co-operation with provincial governments were created (enabling improved co-ordination of provincial efforts abroad), and francophone officials were increasingly admitted, strengthening the cultural dimension of

---

[1] A DFAIT director describing a possible effect of electronic systems automatically searching for and distributing relevant electronic information to hand-held mini-computers of DFAIT-officers irrespective of geographical location. Based on 2001 interviews at DFAIT.

[2] Factors analyzed in the study included service breadth, service depth and customer relationship management. Canada was ranked number one among 21 developed countries worldwide and in fact increased its lead over the closest challengers—Singapore and the United States. See *eGovernment Leadership: High Performance, Maximum Value*, Accenture 2004 (http://www.accenture.ca/xd/xd.asp?xd=insights\insi_egov42.xml)

diplomacy and the focus on multiculturalism (Cooper 1999).[3] These facilitating factors combined with an articulate understanding of the technological challenge at the leadership level in the organization, have provided DFAIT with a unique competitive edge in the process of IT implementation.

The chapter will first introduce the institutional characteristics of the organization. The second part will sketch the technological capacities of the Department. In the third part, IT-effects will be discussed based on empirical data collected at DFAIT between 2001 and 2003. Conclusion relates the empirical evidence to the theoretical framework and assesses the position of DFAIT in relation to the indicators of the magnitude of change of the organizational basis of diplomacy.

4.2 *Institutional Background and Characteristics of DFAIT*

A greater part of its modern history, Canada had the status of a Dominion of the British Empire. It was formally granted independence by the Statute of Westminster passed by the British parliament in 1931. Yet, the British Parliament retained the right to amend the Canadian constitution until 1982 when the current Canadian constitution was adopted. Despite this gradual emergence as an independent nation, Canada still formally remains a constitutional monarchy with the British queen as its head of state.[4] As a consequence of this situation, the Canadian foreign ministry, as Burchill (1993:59) proposed, has been "less confident of itself than has likely been the case elsewhere, because it came into being and continues to operate in a constitutional lacuna." Therefore, when Sir Wilfrid Laurier and his cabinet founded the Canadian foreign ministry on June 1, 1909, it was called the *Department of External Affairs* and not 'foreign' affairs in order not to distort the

---

[3] See also Mackenzie, H. (1998): *Recruiting Tomorrow's Ambassadors: Examination and Selection for the Foreign Service of Canada, 1925–1997* Ottawa: DFAIT (www.dfait-maeci.gc.ca/hist/docs/converted/recruitment-e.htm). In 2001, 83% of DFAIT's 329 managers were fully bilingual (see Performance Report 2001–2002, Ottawa: DFAIT, p. 75), and yet at some Canadian missions abroad, such as for instance the Consulate in Nagoya / Japan in 2000, getting service in French may be problematic (see pt. 1.2.33 of the Audit of the Canadian embassy, Tokyo (including the Canadian Consulates in Nagoya and Fukouka and the Honorary Consul Offices), *DFAIT*, September 2000.

[4] For an overview of the Canadian political system see for instance *Politics of Canada*, Encyclopedia of Political Information at http://www.politicalinformation.net/encyclopedia/Politics_of_Canada.htm

principle of diplomatic unity of the British Empire (Eayrs 1982:96). Reflecting this situation, the point with establishing the Department was according to its first head, Sir Joseph Pope, merely to acquire "a more systematic mode of dealing with what I may term, for want of a better phrase, the *external affairs* of the Dominion",[5] while the actual conduct of Canada's foreign affairs should remain in the hands of the British imperial administration.[6]

From these humble beginnings developed what the former British ambassador, Sir William Hayter, referred to as "one of the highest-powered foreign services in the modern world" (cf. Eayrs 1982:97). In what follows, the institutional characteristics of the Department of Foreign Affairs and International Trade (DFAIT) are introduced focusing on organizational structure, personnel and domestic environment, which, as pointed out in the previous chapter, have implications for how the core organizing principles of diplomacy—hierarchy, secrecy and one-way public communication—are institutionalized and function in various national-administrative contexts.

4.2.1 *Organizational structure*

As in most other countries, the recent decades were characterized by increasing involvement of various parts of the Canadian government in the administration of foreign affairs.[7] This situation was reflected in

[5] "Memorandum for Consideration of the Civil Service Commissioners, 25 May, 1907", Civil Service Commission 1908: minutes of Evidence, I, p. 48, *Sessional Papers of Canada*, XLII, no. 15, 1907–1908 (cf Eayrs 1960:15, italics in the original).

[6] This view derived not merely from Canadian hesitation or reluctance to undermine British supremacy, but reflected also an actual lack of personnel able to conduct foreign affairs. As Earl Grey, the British Governor-General in Canada, complained in a letter to the Colonial Office in London in 1908, "Bryce's [the then British ambassador to Washington who was also in charge of Canada's relations to the U.S.] difficulties in conducting the negotiations [with the U.S. government] have, I am sorry to say, been increased by the chaotic condition of the Administration here *qua* External Affairs. There is no Department, no official through whose hands all matters dealing with external affairs must go. Consequently there is no record, no continuity, no method, no consistency. [...] We have only three men in the [Canadian] Government Service, who have any knowledge of details connected with Canada's foreign relations. One drinks at times, the other has difficulty in expressing his thoughts, and conversation is as difficult as it is to extract an extra tight cork, and the third is the Under-Secretary of State, Pope—a really first class official" (cf. Eayrs 1960:21).

[7] As Under-Secretary of External Affairs Gotlieb argued in 1979: "The fact that there is now a constellation of people and influences drawn from many parts of government, from the Prime Minister down, all having a legitimate stake in the planning,

numerous reorganizations that the federal foreign affairs administration has been going through, most recently in 2004 and 2005.

First steps towards the organizational structure that DFAIT had at the time of fieldwork for the current study (2001–2003) were taken in 1980, when by a decision of Prime Minister Trudeau the foreign service officers of the federal department of Trade and Commerce and the federal department of Employment and Immigration were merged with the Department of External Affairs.[8] The result of this consolidation was a "new" Department of External Affairs created on January 12, 1982, which according to the then Under-Secretary of State for External Affairs Osbaldeston had "a role in government which is essentially different from that of the old Department of External Affairs" (cf. Nossal 1993:37). The difference was related mainly to the fact that the new Department was to fulfill the function of trade promotion and also essentially become the main service provider to other governmental agencies active abroad. This brought about an organizational culture, which was in simplified terms more business-like and less 'diplomatic'.[9] The Department was then eventually renamed to "External Affairs and International Trade" on June 28, 1989. The name "Department of Foreign Affairs and International Trade" was introduced in 1995.

At the time of my fieldwork there were two ministers responsible for DFAIT: the Minister of Foreign Affairs and the Minister of International Trade. Through the two ministers the Department was reporting to the Parliament. In addition to them, the foreign affairs and international trade portfolio was also managed by three Secretaries of State charged with geographical responsibilities.[10] The development aid portfolio was managed by the Minister of International Cooperation responsible for

---

formulation and carrying out of foreign policy, means that External can seldom plan, formulate or carry out policy on its own. Practically everything it does is part of a collective endavour..." (cf. Eayrs 1982:100).

[8] To 750 external affairs officials, 300 were added from the department of Industry, Trade and Commerce, 250 from the department of Employment and Immigration, and 150 from CIDA. Trudau in general did not hide his view of the foreign service as an outmoded and largely obsolete organization, when he for instance held that: "I believe it all goes back to the early days of the telegraph when you needed a dispatch to know what is happening in country A, whereas most of the time now you can read it in a good newspaper" (from a TV interview; cf Eayrs 1982:108).

[9] As Eayrs (1982:106) commented on the merger: "If Big is Beautiful, External appears the winner, its influx from its rivals' ranks confirming its role as a department of supply and services abroad—perhaps at the expense of the traditional foreign office skills of intelligence analysis, policy formulation and the conduct of negotiation."

[10] Secretary of State Asia-Pacific; Secretary of State Latin America and Africa,

the Canadian International Development Agency (CIDA), both placed outside the Department in terms of political responsibility.[11] Several organizations outside the Department's portfolio had accountability relationships with one of the two Ministers.[12]

At headquarters the Department was organized into 8 *business lines*[13] managed by six of the ten Assistant Deputy Ministers (see Table 5 and Appendix 4).

The other four ADMs were in charge of coordinating relations with specific geographic regions. This included Africa and the Middle East (32 missions, 6 satellite offices); the Americas (34 missions, 10 satellite offices); Asia-Pacific (27 missions, 6 satellite offices); and Europe (44 missions, 5 satellite offices). In this way a geographic-functional matrix of coordination and responsibility at the ADM level at the headquarters was in place.

In addition to the headquarters in Ottawa, the Department was serving the citizens through 29 Passport Offices throughout the country and through a network of regional trade commissioners' offices.

Outside Canada, the Department operated through a network of 137 diplomatic missions (embassies, high commissions, consulates) and

---

and Francophonie; and finally Secretary of State Central and Eastern Europe and Middle East.

[11] See Section IV of *DFAIT Report on Plans and Priorities 2002–2003* at http://www.dfait-maeci.gc.ca/department/rpp_2002_2003/section_04-en.asp Unless indicated otherwise, all other information on the departmental structure and resources presented below draws upon this document.

[12] The *Minister of Foreign Affairs* was responsible for the *International Development Research Centre*, which has been helping communities in developing countries to address social, economic and environmental problems; the *International Joint Commission*, a joint Canada-U.S. body that has been managing and protecting lake and river systems on the border between the two countries; and the *International Centre for Human Rights and Democratic Development*, an independent organization that has been promoting human and democratic rights. The *Minister of International Trade* was in charge of the *Canadian Commercial Corporation*, an export sales agency that has been working to expand Canada's international trade; *Export Development Canada*, a financial institution that has been providing trade-related financial services to Canadian exporters and investors; the *Northern Pipeline Agency*, which has been overseeing planning and construction of the Canadian portion of the Alaska Highway Gas Pipeline Project; and the *North American Free Trade Agreement (NAFTA) Secretariat (Canadian Section)*, which has been helping administer the dispute settlement provisions of NAFTA. (Adopted from Section IV of *DFAIT Report on Plans and Priorities 2002–2003* at http://www.dfait-maeci.gc.ca/department/rpp_2002_2003/section_04-en.asp)

[13] For the rationale behind the business line approach in Canadian government see Potter and Bernier (2001).

Table 5: *DFAIT Business Lines and ADM Responsibility 2002* (adopted from DFAIT Report on Plans and Priorities 2002–2003, Section IV)

Assistant Deputy Minister (ADM)	*Business Line Responsibility*
ADM, International Business	International Business Development
ADM, Trade, Economic and Environmental Policy	Trade, Economic and Environmental Policy
ADM, Global and Security Policy	International Security and Cooperation
ADM, Communications, Culture and Policy Planning	Public Diplomacy
ADM, Human Resources	Corporate Services (Human Resources)
ADM, Corporate Services, Passport and Consular Affairs	Assistance to Canadians Abroad
	Corporate Services
	Services to Other Government Departments
	Passport Services

27 satellite offices, supported by 108 honorary consuls in 97 countries (accredited to 192 countries). In the Departmental self-assessment, the missions abroad were seen as providing the possibility of one-stop shopping for federal services and functions in other countries and thereby demonstrating the Department's ability to manage issues horizontally and coordinate a broad range of activities as indicated in the business lines.[14]

4.2.2 *Personnel and organizational culture*

DFAIT employees were divided into Canada-based staff and Locally Engaged Staff at missions abroad. Of the Canada-based staff, approximately 1900 were rotational in 2002. This group included senior executives, foreign service officers (trade commissioners and officers dealing with political/economic issues), management and consular staff, information technology specialists and administrative support personnel. The rest of the Canada-based staff—approximately 2100 officers—were non-rotational and worked both at headquarters (e.g. in Corporate Services,

[14] DFAIT Report on Plans and Priorities 2002–2003, Section IV at http://www.dfait-maeci.gc.ca/department/rpp_2002_2003/section_04-en.asp)

Passport Services or Services to Other Government Departments) and at regional passport offices and trade centers in Canadian provinces. The Locally Engaged Staff—about 4700 employees in 2002—worked as part of various programs at Canadian missions abroad, where their local expertise and local language skills were valued. Approximately one third of these officers were supporting programs of other governmental departments, mostly the immigration program.[15]

Foreign service officers include political/economic officers, immigration officers and trade officers. They are ranked in three basic administrative levels: developmental (non-tenured), working (mid-level) and executive (senior). Each of the levels encompasses three different salary levels, but no formal differences in rank (Jones 2005:44). Canadian diplomats are organized in the Professional Association of Foreign Service Officers (PAFSO), which negotiates most professional issues and organized the first legal strike in Canadian diplomatic history in 2004 with the aim of increasing wages (ibid., p. 43).

The formal requirement to enter the foreign service is a university degree, although most entrants have a post-graduate degree. Applicants must demonstrate proficiency in both English and French (training is provided for 12 months for those in need of improving one of these). Upon entry, there is a 12-month on-the-job training period at headquarters and a two-three year posting abroad. There are evaluations at 18, 36, 48 and 60 months in service and only officers passing these with "fully satisfactory" results can become tenured foreign service officers (ibid., p. 45).

The organizational culture and working ethos of the Departmental workforce has been characterized by informality and flexibility. Such attributes might be seen as being mostly the property of the trade-officials and the flexible business-like culture of the Department certainly did owe a great deal to their presence within the organization. Yet, flexibility and teamwork have also been characterizing the foreign affairs officials ever since the early days of the Department, albeit for different reasons than in the case of the trade officials. As Skilling (1945:274) notes referring to the Department of External Affairs in the 1940s,

> the work of the various divisions naturally overlapped considerably, and members of one division had frequently to be in contact with members of another, and with the Under-Secretary. In view of these facts and the

[15] See ibid.

> smallness of the Department, the distribution of work was flexible and was not rigidly determined by the organization.... The main burden of responsibility concentrated on the Under-Secretary himself, in informal consultation with several of the other top officials of the Department, without the need for a more formal committee on policy...

The fact that the Departmental leadership encouraged such a culture of flexibility and consultation was crucial for further institutionalization of these features at the Department.[16] Maintenance of such an organizational culture was further enabled by the fact that, "the Department, questioning the worth of new management techniques and organization theory, tended to resist administrative innovation" (Eayrs 1982:98).[17] Somewhat paradoxically, this conservatism in its approach to organizational change enabled the Department to keep the culture of flexible mandates and consultation in times of a general rationalization and increased bureaucratization of Canadian governmental agencies. In general, the Department had been relying upon what Cooper (1999:43) had termed *diplomatic improvization*, "geared to operational agility rather than to working within definitive guidelines [...enabling the Department] to deal flexibly with external situations [...]". As early as 1953, the economist B.S. Keirstead referred to these practices at the Department as "an exercise in pragmatic *ad hocery*" (cf. Stairs 2001:17).

The consultative and flexible organizational culture that developed in early decades of the Department's operation remains a strong characteristic feature of its workforce until today. This is reflected in the number of projects, special task forces, virtual teams and task-related temporary divisions continuously formed and re-grouped at

[16] As Eayrs (1982:97) points out, O. D. Skelton, who was the Under-Secretary of State for External Affairs in the period 1925–1941, "imparted to [the Department] through his character and his ways of looking at the world an ethos that outlasted him. [...] Skelton created contours that endured for decades." Maintenance of the organizational culture that emerged under Skelton's leadership was helped by the fact the Department was led in the years 1948–1957 by Skelton's close apprentice—Lester B. Pearson. Hence, despite the post World War II, growth of the Department in terms of the number of organizational units and employees, the organizational culture of internal informality that developed throughout the early decades of its organizational development had remained a strong characteristic of the Department.

[17] In response to a 1963 report by a Royal Commission on Government Organization suggesting private sector inspired changes at the Department, the Departmental spokesman (backed by Pearson, who then was Prime Minister) quoted Viscount Falkland: "When it is not necessary to change, it is necessary not to change" (cf. Eayrs 1982:104).

the Department and throughout the Foreign Service. These tendencies are strengthened by the pervasiveness of client-oriented service ethos throughout the Canadian public administration in recent decades. New Public Management inspired business-like methods are increasingly applied throughout the Canadian government, including DFAIT.[18]

In 2006, in an internal review of the Departmental mandate, the roles of DFAIT were identified as:

- *An interpreter* of international events and trends for the government and for Canadians, recognizing the growing importance of globalization on Canadian's life;
- *An articulator* of a distinctive Canadian foreign policy which expresses Canadians' view of the world in which they wish to live;
- *An integrator* of the government's international agenda and its representation abroad;
- *An advocate* of Canada's values and interests abroad;
- *A provider* of world-class consular and passport services to Canadians;
- A responsible *steward* of public funds charged with delivering common services abroad on behalf of all government departments (cf. Blackwell 2007:49–50).[19]

4.2.3 *Domestic environment*

The domestic environment in which the Department has been operating has been characterized by a federal political structure, a bilingual society and a lack of any clearly defined domestic support constituency for foreign affairs. The organizational changes that the Department has gone through since the 1980s have by and large been attempts to respond to the challenges posed by the domestic environment. As Cooper (1999:42–51) suggests, the move towards adding trade officials who brought in managerial techniques and business-like culture could also be read as an attempt by the Department to gain legitimacy and support in the domestic business community and indeed throughout the

[18] See Savoie (1990, 2004), Potter and Bernier (2001), Clark (2002). For a well-elaborated set of ideas on *New Public Management* including a number of case studies from Australia, New Zealand, Norway and Sweden see the volume edited by Christensen and Lægreid (2001).

[19] In 2006, Adam Blackwell was Director General, Strategy and Services Bureau at DFAIT.

society at large generally skeptical of empowering federal institutions. Similarly, there were ongoing attempts throughout the recent decades at increasing the percentage of francophone officials and at developing frameworks of cooperation with provincial governments (ibid.).

A further challenge for the Department is the system of foreign policy decision-making at the federal level, where the Prime Minister and the Privy Council Office[20] retain the lead in the overall political coordination of Canada's international policy. This creates a situation where the Department continuously needs to jockey for its position as the coordinator of foreign affairs and tries to avoid being merely a service agency for the delivery of governmental policies abroad.[21] Furthermore, the growing internationalization of agenda administered by other governmental agencies has led to a situation where DFAIT also had to fight for retaining particular pieces of the international policy portfolio. As Copeland (2005) holds,

> Encroachment by other government departments into areas previously believed to be the exclusive preserve of the foreign ministry has been going on for years, but there has been no compensatory effort to insert the foreign ministry into domestic debates, to underline its relevance to the security and prosperity of Canadians; there has been no concentrated, sustained or strategic effort to develop a domestic constituency. The days of the foreign ministry cruising on its vestigial prestige and mystique, with its face to the world but its back to Ottawa, have long since passed.

To address this situation, comprehensive IT-supported efforts were initiated to engage the Canadian public in debates and national forums on foreign policy priorities and thereby nurture some form of a broad-based political constituency in the domestic environment (see part 4.6 for more details). Moreover, since the early 1980s, the Department has been encouraging its officers to take on secondments in other govern-

---

[20] The Privy Council Office is "the Prime Minister's department and the Cabinet Secretariat. It is a focal point of action in the Government of Canada's public service. [Its] work requires close and continuous contact with other federal departments and agencies to support their ability to work effectively and to ensure overall consultation and coordination. [They] provide non-partisan advice and support to the government, and leadership, coordination and support to the departments and agencies of the government." (quoted from http://www.pco-bcp.gc.ca/default.asp?Language=E&Page=aboutpco)

[21] DFAIT provides the infrastructure and other support for other governmental departments' activities abroad. See DFAIT Report on Plans and Priorities 2002–2003, Section III—Corporate Services at http://www.dfait-maeci.gc.ca/department/rpp_2002_2003/section_03_04-en.asp?#services

mental departments so as to deepen their expertise in their respective fields of specialization, while at the same time experts from other governmental agencies would be admitted on assignments within the Department (Eayrs 1982:101, Balloch and Angell 1992, Cooper 1999, 2004). DFAIT's integration and cooperation with other governmental agencies has been facilitated by the fact that the Department has been one of the pioneering agencies in the context of the Canadian government in introducing IT in support of its organization and operations, not least as part of the Government On-Line Initiative. This was related to a more general pattern of innovation in foreign affairs administration in Canada focused on involvement of societal actors, which, as Cooper (2004:253) suggests, could be a considered "a model of how foreign policy should be restructured to facilitate competence and credibility and even how the 'foreign' in international activity should be redefined."

The next section provides an overview of the technological capacities available at DFAIT at the time of fieldwork.

4.3 *IT-infrastructure of DFAIT*

Aggressively modernizing its IT infrastructure since the mid-1980s, DFAIT was not caught by surprise when the IT revolution driven by the expansion of the Internet swept the globe in the mid-1990s. The departmental web-site went online on July 26, 1995, way ahead of most other foreign ministries. Since about the same time, the Department has been managing its global activities with the help of a dedicated global Multi-user Integrated Telecommunications Network (MITNET) comprised of leased lines enabling all employees at headquarters and at missions abroad irrespective of geographical location access to the computing platform of the Department SIGNET (Secure Integrated Global Network—a client server network featuring e-mail, organizational intranet and access to the Internet) from their own desktop computers, portable computers or other mobile devices. SIGNET comprised two networks—SIGNET-3 for unclassified/protected traffic, which in 1999 supported over 8000 PCs with local area networks at HQ and at 86 missions and stand-alone PCs at 75 smaller missions; and SIGNET-C4 carrying classified and protected/sensitive traffic. Besides daily electronic management of documents and e-mail, SIGNET-3 was used for accounting, personnel- and property management on a global scale, as well as for telephony. Furthermore, through the so called SIGNET MetaFrame, staff at embassies could access about 118 databases containing over 100 million pages with foreign affairs related information on a 24/7 basis (including for instance Dow Jones, Oxford Analytica, The Economist, all major newspapers and other sources).[22] Both MITNET and SIGNET provided the main communication backbone for all government of Canada operations overseas,[23] and other Canadian governmental agencies active abroad hence did not need to develop their own communication infrastructures.[24] In 2005, about 33% of SIGNET users were located at DFAIT headquarters in Ottawa and 67% at Canadian missions abroad (of these about a half were employees of federal departments other than DFAIT).[25]

Concentrating the maintenance and operation of governmental communications' systems at DFAIT was cost-effective from a government-wide perspective, but required the commitment of a fairly extensive staff and resources at DFAIT to keep the system operational at all times. It is hence not surprising that the Information Management and Technology Bureau (SXD) of DFAIT was with its 425 full-time employees and 100 consultants the largest bureau of the Department at the time of fieldwork. As Smith and Sutherland (2002) point out, this has in turn had implications in the form of greater awareness of IT and positive attitudes to using IT applications throughout the organization. Such attitudes were fostered also by the senior management in the Department consciously cultivating a sense of adventure about technology within the organization and providing officers with great leeway in experimenting with new tools such as the Internet.[26] All in all, as Evan Potter (2002) points out, the technological capabilities of DFAIT were an envy of most foreign ministries around the world at the

[22] In 2001, the Department used CAD 1.3 million to cover subscription fees to these resources. (Based on 2001 interview information.)

[23] 7.5% of its overall annual budget or about CAD 100 million were spent by the Department in 2001 on the IT infrastructure. (Based on 2001 interview information.)

[24] The U.S. government, for instance, maintains several information infrastructures, with each agency active overseas using its own system. This creates inefficiencies in particular in small missions, where an officer often has to switch between several terminals and security systems while working.

[25] *IM/IT Needs Analysis Study*. Office of the Inspector General, Evaluation Division, DFAIT, November 2005, p. 6.

[26] Ibid., p. 166. A great advocate of using information technologies to support work of Canadian diplomats, the Deputy Minister of Foreign Affairs 1993–97, Gordon S. Smith, is a leading example of IT-savvy enlightened leadership at DFAIT.

time when the fieldwork for the current study was conduced. Indeed, the results of a survey conducted through focus groups, in-depth interviews and mission-teleconferences with over 200 DFAIT managers and other experienced staff in mid-2005 had shown an almost unanimously shared perception that the department "had now 'gotten it right' with a common infrastructure that is pervasive, stable and reliable."[27]

What then have been the effects of IT on the core organizing principles of diplomacy embedded in DFAIT?

4.4 *IT-effects on Hierarchy*

As elaborated above, effects of IT on hierarchy in a foreign service can be operationalized in two ways—1) bureaucratic hierarchy and 2) hierarchy in the relations between headquarters and missions abroad.

4.4.1 *Challenges to bureaucratic hierarchy*

Pervasiveness and ubiquity of e-mail communication and Internet access throughout the Canadian foreign service have, as one of the senior officers put it in 2003, "created conditions for short-circuiting the hierarchical lines of authority at the Department". Indeed, as another senior officer observed, challenges to the bureaucratic hierarchy were coming primarily from junior officers who were not socialized into the bureaucratic habits and used IT to communicate directly with senior officers. The latter for the most part did not discourage this behavior and communicated back, so that formal hierarchical lines of authority often did not correspond with the flat communication patterns in day-to-day operations of the Department. However, as different technologies were becoming available, their implementation and use by the Department has always been limited by adherence to bureaucratic standards and/or the development of new procedures that enabled the application of such standards in the electronic environment. This was demonstrated at two levels: 1) internal electronic communication, and 2) filing of documents in electronic archives.

---

[27] *IM/IT Needs Analysis Study*. Office of the Inspector General, Evaluation Division, DFAIT, November 2005, p. 9.

4.4.1.1 *Internal electronic communication and bureaucratic hierarchy*

The high level of connectedness and ease of communication introduced by the SIGNET system has led to unprecedented increases in the volume of information flow within the Department (30 million messages sent in 1998,[28] estimated 60 million messages sent in 2001)[29] and an unprecedented pressure upon the higher ranks as the average number of e-mails received by a director per day was 83 in 2001.[30] As a result, many senior officers came to change their attitude to answering e-mail messages—having 300–400 unanswered e-mails in the In-box came to be perceived as a standard situation by many.[31] An internal information flow analysis conducted by the tech-support staff of the Department in 2000 indicated that a large number of the messages circulated at the Department were broadcasts. It was not unusual that these had to do with semi-professional issues or were of personal nature such as announcing the sale of a car and the like, and it was not unusual that important messages were overlooked by officers in the loads of internal spam. The Department had therefore decided to limit the number of broadcasts in a hierarchical manner where only directors and those above them in the bureaucracy would be allowed to send them. E-mail technology was appropriated to the existing bureaucratic structure in order to exercise a higher degree of control on its potential to paralyze the organization by information overload.

Another indicator that the bureaucratic procedures were not undermined by IT was the fact that despite increasing budgets on technology throughout the 1990s, the use of paper had not dropped. On the contrary, it has increased by 400% in the period 1993–1999.[32] Largely, this can be attributed to the employees' unwillingness to go over to fully electronic transactions, which in turn was caused by the lack of a clear set of authoritative Department-wide procedures and routines for such transactions. As the above mentioned internal report states:

---

[28] *DFAIT Information Management Strategic Plan 2000/01–2002/03*, internal DFAIT report, 6 July, 1999, p. 19.

[29] Based on 2001 interviews.

[30] Based on 2001 interviews.

[31] The excessive amount of e-mail may, however, also be related to the fact that only a "few [employees] used the 'message filtering' feature" in the Outlook 98 program in 2001. See *Audit of Desktop Tools*, DFAIT, May 7, 2001, p. 18.

[32] *DFAIT Information Management Strategic Plan 2000/01–2002/03*, internal DFAIT report, 6 July, 1999, p. 21.

> Much paper generation results from isolated solutions to particular problems: e-mail and local printing to replace telegram distribution, Jetform "print on demand" to replace central printing and stocking of forms, applications which require manual re-entry of information, sometimes several times. Whatever the origin of the "paper problem", it points in many cases to ineffective implementation of technology, e.g. reproduce a traditional process, without examining whether it continues to be valid. Interviewees commented on the many requirements for signature (e.g. in appropriating leave), which automatically requires a paper transaction.[33]

In addition to the inefficient coordination between implementation of new IT and development of new routines and procedures for their use, the reason why bureaucratic hierarchy had not been undermined was also related to the very nature of the overloaded informational environment that IT had created. Bureaucratic hierarchy became a tool enabling the organization to solve the *paradox of plenty*. This notion elaborated by Nye (2002:67) drawing upon the work of Simon (1998), describes a situation where due to advanced IT, actors have access to more information, but may be less informed. Although not explicitly stated by Nye, one might argue that the term is related to the notion of limited *absorptive capacity* (Cohen and Levinthal 1990). It conceptualizes the fact that with full freedom of access to information but lacking reliable informational filters, organizations may not be in the position to extract relevant information and decision-making may be severely hampered or entirely paralyzed. In the above example of broadcasts bureaucratic hierarchy enabled the limitation of access to such an e-mail function to the top echelons of the organization and thereby reduced the flow of information. This was a decision implemented by the IT support units in consultation with the senior levels in the Department, and therefore more or less a top-down approach to using hierarchy for structuring the internal information flows. On the other hand, as a senior diplomat observed, junior officers were also motivated to use the hierarchical system to their own benefit as gaining the approval by an officer above would provide their initiatives, e-mails or documents with higher level of *attention* and *credibility* in the informationally overloaded internal environment.[34] There was hence also a bottom-up set of reasons for keeping the bureaucratic hierarchy intact in electronic transactions.

---

[33] Ibid.

[34] Based on 2003 interviews.

4.4.1.2 *Electronic archives and bureaucratic hierarchy*

In 2001, there were a number of technological solutions available for electronic management of documents. There was an attempt to introduce a shared electronic archive called CATS,[35] which would feature all unclassified documents and provide a broad access to these documents for all SIGNET users, that is throughout the entire foreign service. Theoretically, such a solution should have allowed for a cross-unit and foreign service wide access to unclassified documents and should hence undermine the standardized bureaucratic stovepipes horizontally, but also vertical differences in access to departmental information, and hence potentially undermine the bureaucratic hierarchy in the organization. Yet, as practice has shown, only about 0.5 % out of all unclassified traffic exchanged within the Department per year were actually filed in the CATS electronic archive in 2000.[36] As a 1999 internal DFAIT report pointed out:

> Employees at DFAIT feel that there is a significant loss of corporate memory that must be addressed by the Department. Aside from the direct impact on productivity, fear of loss contributes to excessive printing of documents and redundant use of storage and other IT infrastructure.
>
> [...S]taff are unable to find the information they need quickly and accurately, and [...] information chains, from data creation to use, are not always identified and managed. The challenge here, however, is that the Department does not always know what information it has or understand the processes that create information and transform it into a useful, reusable asset.
>
> During the 1980s, as DFAIT began to move from paper- to electronic-based files and records, the policies were not necessarily updated to reflect the change in procedures. Although this was eventually recognized and policies were established for certain procedures, such as consistently archiving e-mail records, there was still minimal compliance. Currently out of 30,000,000 messages sent within DFAIT each year, only 90,000 are filed in CATS. The result is a significant void in valuable business intelligence, business assets and records that could have been used for current and future reference. This point is critical given the fact that most staff are knowledge workers, completely dependent on access to information to perform their jobs."[37]

---

[35] CATS is an acronym for Corporate Automated Text Storage.

[36] *Audit of Records Management*, Audit Division, DFAIT, May 2001, p. 16. As the report further points out, "If the information is not in CATS (or any other shared system), employees see no benefit in bothering with the system, so new material is not entered, and the cycle continues" (ibid.).

[37] *DFAIT Information Management Strategic Plan 2000/01–2002/03*, internal DFAIT report, 6 July, 1999, pp. 19–20.

Usage of CATS was low not only in terms of the volume of documents filed in the system, but also in terms of the actual number of officers using the application. As a 2001 survey of 644 respondents throughout the foreign service had shown, only 37% had been active users of CATS (at headquarters, the number was 47% of employees, and at missions, it was 30% of employees).[38]

There may be several explanations for the low usage of CATS. One of the factors is lack of training because in 2001, only 25% of headquarters' staff and 37% of mission staff had received training in using CATS.[39] Related to this is the factor of individual employees' forgetting to file documents in CATS.[40] Another explanation was related to secrecy norms inherent in the foreign service and is discussed below in the part 4.5 below. Finally, the low use also had to do with unclear status of e-mail messages and working versions of electronically exchanged documents. Given the lack of authoritative routines and procedures for archiving electronic information, many officers had problems determining what should and should not be kept in the departmental electronic archive. As a senior officer observed:

> There are uncertainties as to what information or document is actually worthy of being kept in the archives. Only the final versions of documents? Or also preliminary versions of documents? And what about comments by officers involved in the production of a document? Furthermore, there is also the unclear nature of e-mail—to what extent is it official communication to be kept in the archive and to what extent should short messages just be regarded more or less as oral communication?[41]

In addition to the ambiguities related to electronic production of documents and communication, the low usage of the CATS electronic archive could also be related to what could be labeled a *bureaucratic mindset* within the Department. As one of the FSOs I interviewed in 2003 pointed out:

---

[38] *Audit of Desktop Tools*, DFAIT, May 7, 2001, pp. 5–6.

[39] Ibid., p. 7.

[40] This was cited as one of the main reasons for the low usage of CATS in the *Audit of Records Management*, DFAIT, May 2001, p. 1.

[41] Based on 2003 interview information. The 2001 *Audit of Records Management* (p. 31) does provide some form of guidance in relation to this issue: "Technically, there is no such thing as a 'working file', from a records management standpoint, if the 'working file' contains the original or only a copy of a document. If a document is created in the course of doing business, it is a record. More pragmatically, early drafts and rough notes that have not been sent to anyone else, can be deemed 'transitory' and may be destroyed, but that is virtually the only exception".

> The budget of my division is unclassified. If we file it in CATS, then anyone in the Department can see what resources we have and that is clearly a disadvantage.

Apparently, transparency and openness that on the one hand would allow for faster access to information throughout the Department and thereby improved use of overall knowledge resources of the diplomatic workforce on the other hand also involved a disadvantage for individual organizational units in the internal turf-battles over financial resources. This is an example of an *institutional collision* between the diplomatic norms and the bureaucratic norms. The former lead foreign service officers to produce high-quality diplomatic analyses for which they need access to as much information available throughout the Department as possible and therefore to demand of as broad access to electronic information throughout the Department as possible. The bureaucratic norms, on the other hand, limit their willingness to share information for fear of disadvantages in the internal budget redistribution processes so central to the internal dynamics of any bureaucracy.

In any event, instead of filing documents in the central electronic archive, officers and organizational units started to create their mini-archives on their own PCs or at best in the database of their respective unit. In an internal report, this situation was described as

> the ideal environment for 'data-hoarding' [...], where individuals hold much of the corporate information on their 'personal' computers. The results are duplication, multiple yet different versions, and lack of a single, authoritative copy of a document.[42]

In another report, this situation was analyzed in the following way:

> Most groups [within the Department] can find the documents they want, when they want, although the efficiency of this process is imperfect and often time-consuming. The same applies to individuals who manage their own records directly. This type of 'stovepipe' situation, where individuals or units manage their own records, typically leads to information isolation. Almost by definition, the information base these employees are working with is limited. [...] They have the information they need, generally, and are not aware of the wealth of information that could be available to them, or others, if the Department's records were properly indexed and universally accessible.[43]

---

[42] *DFAIT Information Management Strategic Plan 2000/01–2002/03*, internal DFAIT report, 6 July, 1999, p. 20.

[43] *Audit of Records Management*, Audit Division, DFAIT, May 2001, pp. 15–16, italics added.

Given the 'stovepipe situation' and the individualized procedures for records management, DFAIT was "misplac[ing] knowledge every time an employee [was] re-assigned."[44] What complicated the situation further was the fact that even in these personal mini-archives on own PCs, the data-hoarding officers themselves had often difficulty in differentiating clearly between personal and official communication, internal and external communication etc.[45]

Overall, these practices led to inefficiencies and low capitalization on the investments made in the technological infrastructure of electronic archives. Illustrating this, an incident was quoted by the internal audit team regarding a particular agenda that was negotiated by the Department with another government in the early 1990s: "When the topic recently came up again, no record of the substance of these original discussions could be found by DFAIT (or by the foreign government for that matter). As a result, both parties agreed not to spend more time looking for the documents and started over."[46]

To improve these inefficiencies, the Department has in 2003 started to introduce a system for the electronic management of documents called *InfoBank*.[47] The system had 2 main goals: 1) improve the maintenance of corporate memory; 2) improve responsiveness and flexibility in relation to the access-to-information requests.[48] Incentives for introducing the new system and the overall vision were summarized by an internal audit team in 2001, who held that

> a more comprehensive change to the Department's information management practices, that, in effect, starts anew should be considered. The promise of such a solution would be that there is a place for every record—paper and electronic—in the corporate system and that staff would store all records in the right place the first time. At any time, the location of all records would be known.[49]

---

[44] *DFAIT Information Management Strategic Plan 2000/01–2002/03*, internal DFAIT report, 6 July, 1999, pp. 20–21.

[45] Based on 2001 interview information. Similar observations of individualized practices for storing electronic information were made in *Audit of Records Management*, Audit Division, DFAIT, May 2001; and also in *Audit of Desktop Tools*, DFAIT, May 7, 2001

[46] *Audit of Records Management*, Audit Division, DFAIT, May 2001, p. 18.

[47] InfoBank is a DFAIT label for the Records, Document and Information Management System (RDIMS) a Government of Canada Shared System developed for use by all federal departments.

[48] Based on 2003 interviews.

[49] *Audit of Records Management*, Audit Division, DFAIT, May 2001, p. 32.

Development of the system and its rollout throughout the foreign service was not managed by technical support staff but by experienced Foreign Service Officers (FSOs). As one of the informants pointed out in 2003, this was done mainly for two reasons. Firstly, the FSOs were better in position to judge what features of the system would be of use for the daily work of a diplomat than their tech-support colleagues. Secondly, having experienced FSOs in charge of such a comprehensive project involving the whole foreign service would also ensure that the project would receive appropriate attention at the senior level within the Department. Finally, one can observe, that the direct participation of experienced FSOs in the development of the InfoBank software and its architecture ensured also that the system would feature characteristics conducive to the norms and practices of diplomatic work.

The fundamental systemic difference that was introduced with the InfoBank was that access rights were attached to each and every document (including unclassified documents) and the rights were determined by the officer who filed the document. Before saving it in the InfoBank for the first time, the interface would show an electronic form, where the officer needed to determine classification level and access rights by ticking off fields. To avoid laborious filling out of this form every time a document was saved, there was a function enabling the creation of a template for the level of classification and access rights (e.g. personal access only, access to colleagues in the same unit and the supervisor, or other kinds). The document or e-mail was then saved into the database and was searchable from any machine connected to the SIGNET system. Access to the document was available based on the access criteria attached to each document. Hence, while one could find documents and thus know they were available, if one was not authorized to view a particular document, one needed to ask for access from the officer who filed it. The notion that this represents the enactment of the *need-to-know* principle in the electronic environment is discussed in more detail below in part 4.5.

What is interesting to note here was the *potential double effect of the InfoBank system on the maintenance of boundaries between bureaucratic organizational units.* On the one hand, the system allowed for very clear maintenance of the bureaucratic boundaries between organizational units by enabling the standard setting of access rights to include members of a specific organizational unit and/or its sub-units. In such a case the informational boundaries would copy the existing bureaucratic organizational structure and hierarchy. Yet the way the system was constructed, it also allowed for the creation of sharing systems featuring organizational units and/or individuals involved in the work on specific agenda and not necessarily located in the same organizational unit. In such a case, formal bureaucratic boundaries and hierarchies within and between organizational units would be supplemented by a set of issue- and task-specific informational boundaries. Firstly, this might have the form of cross-unit informational integration of exiting units.[50] Secondly, it would potentially also involve the creation of informationally integrated cross-unit cross-hierarchy groups and task-forces within the Department, and hence also the development of an alternative informational structure, not entirely consistent with the formal bureaucratic structure. The flexibility in cross-unit cooperation and the consultative culture that have been characteristic features of the Department since its foundation might come into play here and reinforce this side of the connective potential that the InfoBank system provided. On the other hand, the nature of the bureaucratic procedures in place might limit such alternative informational structures from arising.

In an ideal Weberian bureaucracy a set of documents and files constitutes one of the essential characteristics of a bureau. It is one of the means in which individual units in a bureaucracy delimit their boundaries. The informational sharing systems and integrated databases enabled by InfoBank challenged these boundaries in terms of a) breadth of informational resources used in the production of documents; b) breadth of consultation (the ease of access to electronic documents enables a higher number of relevant officers to get involved in consultation and reviewing drafts); and c) location of document production (documents might be produced in different locations than the bureau carrying responsibility). These factors should theoretically contribute to improved quality of outputs produced by DFAIT. At the time of fieldwork, the system was in its pilot stage and further investigations would be needed to support such a contention. As an internal DFAIT survey had shown in 2005,[51] there were a number of challenges regarding the

[50] According to interview information, in late 2003, one considered developing a Northern Asia regional approach to integrating Canadian diplomatic informational resources related to the region. This would involve creation of a sharing system within the InfoBank between the China-Mongolia Division at DFAIT and the Canadian embassies in Beijing, Bangkok, Seoul and possibly others in the region, as well as possibly including the Japan division and the Canadian embassy in Tokyo. Academics and NGO actors were to be included on a consultational basis using e-mail—not having access but included in ad-hoc information sharing.

[51] *IM/IT Needs Analysis Study*, 2005.

application of InfoBank, mainly relating to the laborious and time-consuming tasks that needed to be performed by line officers creating and filing documents. As the above mentioned internal report states:

> Participants in several focus groups alluded to the need for tools that automatically index emails, documents, Web pages, notes and other useful information so that they could find and retrieve them immediately when needed. Individuals who had found the free Google Desktop tool very useful in this regard at home wondered why there was nothing similar on their SIGNET desktops. It was pointed out that InfoBank was not a viable answer to this need because of the time required to individually enter each document and email and the fact that the majority of messages are not seen at the time as being of enduring organizational value ensures that individuals will realistically archive only a fraction of their daily work. Instant access to the totality of one s previous work was seen as a major contributor to personal productivity and effectiveness, as well as a partial solution to the problem of fragmented information stores.[52]

As numerous participants in the survey reported, the 5 to 10 second delay needed to store an e-mail in InfoBank led to great productivity losses and resulted in 95–99% of e-mails not being archived. Moreover, as the Ministers' offices were not to be connected to the system, many officers perceived that the system would have inherent gaps of corporate memory.[53] Of the 765 officers at headquarters who had completed InfoBank training by November 2005, 40% reported that they were saving at least some documents through InfoBank, but only 20–25% of organizational divisions at headquarters could be said to have implemented the system into daily work patterns.[54] To address this situation, the Chief Information Officer at DFAIT has established the so called InfoBank Management Board in 2005. Featuring officers from the headquarters and from the missions, it has been advising on functional changes, implementation approaches and practices in relation to the InfoBank system.[55]

In sum, although hierarchy remained an organizing principle ensuring political accountability at DFAIT, IT solutions enabling informational integration across formal bureaucratic boundaries of units and flexible involvement of external actors in virtual team cooperation. Still, chal-

---

[52] *IM/IT Needs Analysis Study*, p. 25.
[53] *IM/IT Needs Analysis Study*, p. 27.
[54] Ibid. p. 29.
[55] Ibid., p. 75.

lenges remained pervasive regarding the ability of the Department to organize information in shared databases providing officers with timely access to relevant information. Without such an effective and shared corporate memory, DFAIT recognized that its credibility as the source of authoritative information on foreign affairs would be questioned.[56]

4.4.2 *Challenges to the hierarchy between headquarters and the missions abroad*

As the SIGNET and MITNET networks have enabled an unprecedented connectedness and flexibility of DFAIT operations on a global scale, new opportunities arose for horizontal cooperation and coordination. Informal and formal virtual teams had become the standard operating procedure for DFAIT and its missions and tasks were often being assigned regardless of geographic location, which provided opportunities to integrate expertise throughout the organization (Smith and Sutherland 2002:169).[57] The fact that the searchable database of DFAIT employees available at the DFAIT web-site featured all employees including those at missions abroad could be considered an indicator that DFAIT indeed did present itself as *a single global organization*, and not as a collection of organizations with headquarters at the center. When conducting a search, an employee would be shown with an acronym of his/her organizational unit, a phone number, and a small graph indicating where in the hierarchy the employee is located, including the information about the line of senior officers under which s/he is administratively placed. The format would be the same for officers at headquarters and for officers at missions abroad—not knowing the acronyms, one had a hard time finding out whether the employee was physically located in Ottawa or somewhere else around the globe.[58]

An important factor in the informational integration between headquarters and missions is the *Intranet*. Although not entirely without problems, officers at Canadian missions operate more or less in the same user-environment as officers in Ottawa.[59] Complementing the

---

[56] *IM/IT Needs Analysis Study*, p. 23.

[57] Several of my informants have confirmed this view.

[58] The acronyms could be figured out once one clicked on the 'advanced search' button, where all the units including embassies were shown with full name and the acronyms.

[59] Due to limited bandwidth, some of the missions have been experiencing rather slow access to Intranet applications, e.g. downloading a travel claims form might take as long as 15 minutes. Hence, as the mission in Buenos Aires reported in 2005, Intranet

Intranet-connectivity, there has been a wide-spread practice at DFAIT of sharing electronic information between headquarters and missions. In 2001, 81% of 644 respondents to an internal survey indicated that information sharing is very important in their jobs. 78% of the respondents reported that they typically share files by e-mail and through shared drives with colleagues at missions and 73% reported the same about colleagues at headquarters.[60] Methods of file sharing are summarized in Table 6.

Table 6: *File sharing methods at DFAIT in 2001* (adopted from *Audit of Desktop Tools*, DFAIT, May 7, 2001, p. 13)

File sharing methods (based on 644 respondents)	*HQ staff*	*Mission Staff*
E-mail	89%	79%
Shared Drives	81%	86%
Over the Web	29%	26%
Public Folders	27%	44%
CATS (electronic archive)	20%	19%

Given the relatively high level of information sharing and all the extra activity and time this requires, one had been trying to find ways to facilitate the processes. This included the above mentioned attempts to integrate databases, electronic archives and other informational resources of geographic units at headquarters and the respective embassies around the world with the help of the *InfoBank* project. Geographic units at headquarters in charge of relations with large countries often did not have more than about a tenth of the staff present at the Canadian embassy in the capital of that country. In such cases, due to the connectedness provided by IT, embassies conducted most of the strategy development and analysis and even tasked the respective geographic unit at headquarters when necessary. As one of the senior diplomats put it, the value added by the geographic units at headquarters in such situations is that of coordination on the ground in Canada, i.e. within the Department, but also within the Government more generally and

applications "are either not used at all or, if used, they take so long that they either time-out, do not function regardless of how persistent the user is, or take so long that it has an enormous negative impact on productivity". See *Audit of the Management of the Intranet*. FAC, Office of the Inspector General, Audit Division, January 2006, p. 23.

[60] *Audit of Desktop Tools*, DFAIT, May 7, 2001, p. 12.

indeed in relation to Canadian actors involved with the particular foreign country in question.[61]

The Japan Division at headquarters provided a good example of such a situation. In late 2003, they had 12 staff, while the Canadian embassy in Tokyo had about 150 employees. When the InfoBank project was pilot-tested, the electronic archives of the Tokyo embassy and of the Japan Division were integrated. This was particularly helpful because there is a 13-hour time difference between Ottawa and Tokyo, and very often officers on either side would have to wait a whole working day to receive necessary documents that they requested from their colleagues by e-mail. Due to the informational integration provided by InfoBank, the formerly hierarchical relationship between geographical units at headquarters and embassies was in some cases reversed, while in other cases (in particular in cases of the one-man "laptop-embassies" or other small missions), the relationship remains hierarchical in the sense that geographical units at headquarters carry out analysis and send instructions to the Canadian representatives on the ground.

While IT-supported integration between missions and headquarters was in general perceived as a positive development, a 2005 survey had also revealed perceptions of negative aspects related to this. One of the key issues is excessive information overload produced by headquarters overflowing electronic mailboxes of embassy staff. As the internal report states:

> Mission personnel pointed out that the reason the Departments invest so much in keeping them abroad is so that they can interact directly with foreign officials, businesses and society; not to sit in their offices clearing their inboxes. Many felt that the mundane demands of dealing with email overload detracted them from their ability to focus at the strategic level. As one officer pointed out, 'it controls us'.[62]

Moreover, the electronic informational overload generated by headquarters led to a situation, when a busy foreign service officer at a mission abroad, whose time can cost the taxpayer more than CAD 200 an hour may in fact spend as much as half a day or more every month culling messages in order to clean his/her mailbox.[63] To use time effectively at a mission abroad (in particular in cities like London involving long

---

[61] Based on 2003 interviews.
[62] *IM/IT Needs Analysis Study*. p. 21.
[63] *IM/IT Needs Analysis Study*. pp. 22–23.

commutes to various assignments), virtually all Canada-based staff stationed abroad were hence also increasingly equipped with electronic devices enabling mobile work, where the digital personal assistant BlackBerry was the most preferred system used in 2005.[64] The level of ubiquity and attachment of DFAIT officers to these devices were expressed by one of the senior officers interviewed in 2003, who pointed out that their nick-name at DFAIT was "CrackBerries" (invoking the addiction to the drug known as 'crack'). Yet, a problem hampering effectiveness of mobile officers was that BlackBerries did not enable opening attachments. Moreover, DFAIT lacked an organizational strategy including policies supporting its increasingly mobile workforce.[65]

In sum, IT-supported systems have enabled an informational integration and cooperation between units at headquarters and missions abroad, which in some cases enabled the relocation of policy analysis and decision-making to embassies and raised the issue of the continued relevance of different working styles at home and abroad.

4.5 *IT-effects on Secrecy*

Secrecy was operationalized along two dimensions—approach to electronic risks and policies of internal access to information. Effects of IT on secrecy at DFAIT were studied in relation to these two dimensions.

4.5.1 *Approach to electronic risks*

SIGNET was originally a "closed" network, which provided DFAIT with effective means of *risk avoidance*. However, pressure from system users and from other agencies of the Canadian government to make it easier to access the system have led to more openness.[66] At the time of fieldwork, DFAIT was using an unclassified and a classified electronic network—SIGNET-D and SIGNET-C respectively. Embassies were connected to both, and most of the larger missions featured classified local area networks. The Department had recognized that the IT-enabled opportunities to improve information accessibility and overall knowl-

---

[64] Ibid., p. 39.
[65] Ibid., p. 65.
[66] *Information Management Strategic Plan*, p. 18.

edge management (that were considered strategic goals at DFAIT)[67] would remain largely unused if paper-based security standards would be expected in the electronic environment.

In the light of the development of DFAIT's separate information systems and of the specific information security procedures attached to it, challenges arise in the communication with other governmental agencies active internationally. A case in point discussed in this connection during my visit to DFAIT in November 2003 was the presence at DFAIT of seconded experts from the Justice Department. While located at DFAIT these experts were working on legal aspects of international trade issues. Yet given the fact that informational jurisdictions of Canadian federal departments are legally separated (as in most other modern bureaucracies), there were two intertwined practical problems, which have complicated the experts' work significantly. On the one hand the documents that these lawyers produced were considered the property of the Justice Department and hence could not be stored in DFAIT's databases. Secondly, there were no links between the information systems of DFAIT and the Justice Department, which was substantially slowing down the work of the legal experts, who were in constant need of accessing various legal documents in the database of their own department. There was hence a sense expressed by most of the respondents that secrecy concerns might be hampering effective use of DFAIT's informational resources and thus be a hindrance in effective diplomatic work. DFAIT states this clearly, when it points out that

> there is an inherent tension between security on the one hand and affordability and functionality on the other: no system can offer both complete assurance of security and all of the functionality that its users require.[68]

Efforts to ensure functionality and flexibility have increasingly been gaining ground and security standards had to be adapted to this development. The senior management of the Department played an important role in facilitating this development. For instance, as a senior officer reported to me in 2001, the then foreign minister Manley was determined to receive his e-mail on a hand-held wireless device despite the fact that information security experts of the Department initially considered this an information security hazard. The minister

[67] Ibid., p. 10.
[68] Ibid., p. 18.

eventually had his way and information security staff had to come up with a technological solution that would ensure at least some level of communication security. At the time of writing, e-mail communication through wireless hand-held devices had become a normal working procedure at DFAIT, and the overall approach to electronic risks matched the notion of *risk management* as demonstrated by broadly available access to the Internet and e-mail communication from stationary workstations and mobile devices throughout the organization.

4.5.2 *Policies of internal access to unclassified information*

Due to increasing integration with domestic governmental departments and increasing involvement with non-governmental actors in recent decades, there has been a growth in the volume of the non-traditional diplomatic agenda administered by DFAIT. Due to this development, the overall percentage of classified information administered by DFAIT had been decreasing. Nevertheless, there remained a certain amount of information and agenda that needed to be classified. Information security concerns and secrecy as a professional norm hence remained important factors that DFAIT was taking into consideration in relation to the development of its information infrastructure.

Increasing reliance of DFAIT on IT-supported communication had challenged internal practices related to information security. In the 1980s, prior to the advent of advanced IT, each embassy had an officer in charge of communications with external actors and the HQ. This officer was specially trained in the procedures of classified communication.[69] Foreign service officers have hence readily assigned documents classified status, because this did not represent any additional effort for them personally (the communications officer would take care of the procedures required for communicating the classified document). What is more, documents would in this way be bestowed with higher status and value in the eyes of the recipients. This model was no longer in use at the time of fieldwork. Instead, each individual officer was in charge of communicating documents, be they classified or not. Application of security routines for communications at embassies and at HQ had

[69] As a consequence of the connecting of missions abroad to SIGNET, most of these communications staff (about 130 employees) have been relocated back to the headquarters in Ottawa by 1995 (*Client/Server Case Study*..., pt. 4).

moved from the specially trained officers, towards all foreign service officers authorized to access classified information. Given the specialized procedures that need to be followed in communicating classified information, there was a tendency among the officers to consciously assign documents low classification level or none at all in order to communicate them more quickly using procedures for unclassified communication. This lead to what one of the informants had called "*watering down of classification awareness*" among foreign service officers.[70]

This shift had led to several situations, where classified information has been stored in the unclassified electronic archives (CATS) accessible throughout the foreign service (including embassies through SIGNET-D). Although, as pointed out above, only a fraction of the actual volume of electronic communication is being stored in CATS, this is still perceived as representing a security risk. As several informants pointed out, these risks are related mainly to locally engaged staff at missions abroad, who all have access to CATS on a par with Canada-based staff.[71] Indeed, several situations were registered by DFAIT's information security officers when locally engaged staff at embassies (notably in one of the major Asian capitals) were detected accessing files in CATS and as those moves were reconstructed, documents that were accessed did in aggregate contain sensitive or classified information. As one of the informants pointed out:

> We do some monitoring but we cannot aspire to uncover everything. If someone is good at erasing traces of being in the system, or can hide well, we may not find out for years. Earlier, someone would have to break into the building, pass the guards and then break into safes to steal classified information. Now all they need is a laptop with an internet connection somewhere in a basement on the other side of the globe. We've had hundreds of years to figure out how to lock up pieces of paper. But we still have not figured out ways of completely securely locking up a server. DFAIT has become tremendously vulnerable, and in recent years there has been a proliferation of threats to information security.[72]

[70] Based on 2003 interviews.

[71] This was also related to the relatively weakly institutionalized procedures for publishing documents on the *Intranet*, where locally engaged staff without the appropriate security clearance could access personal information or property descriptions. See *Audit of the Management of the Intranet*. Office of the Inspector General, Audit Division, January 2006, p. 24.

[72] Based on 2003 interviews.

One of the emerging security issues that concern information management procedures rather than the technological infrastructure were the *chains of e-mails*. As several of my respondents have reported, it has been a steady occurrence that e-mail chains containing communication around a specific issue are routinely attached and copied to various recipients within and outside the Department. Over time and dozens of e-mails later, the chain may include information that is classified or sensitive if aggregated or if read in the particular issue-context that the e-mails in question concern. As a senior officer pointed out to me:

> Say somebody starts to communicate [by e-mail] about regulations on the imports of Swiss cheese. Harmless, it seems. The e-mail exchange goes on to include other actors and previous e-mails are always attached, and as the issue discussed gradually develops, suddenly 200 e-mails later, you look back and realize, oops, this is classified info, we shouldn't really be talking about this freely.[73]

The ease with which e-mail communication enables attachment of the previous communication and distribution to a broad range of actors produces a level of ambiguity when it comes to information security. Although an individual officer may be sending an unclassified e-mail, in combination with the attached chain of e-mail messages, it may in fact be sensitive or classified. In response to this situation, DFAIT implemented specific procedures reintroducing the *need-to-know principle* in the electronic environment.

As in most other foreign ministries, the need to know principle has been a characteristic feature of DFAIT's policy of internal access to unclassified information. The formal procedures have an informal underpinning in what might be termed a need to know culture. As a senior officer pointed out:

> The need to know principle is happening on a personal basis. Those are the meetings that happen after 6 o'clock. A couple of people sit down, the doors close, and nobody outside knows what is really going on.

Although IT obviously does not challenge such informal practices, there have been technological developments that have seemed to challenge the practice of applying the formal need to know principle in regulating access to unclassified documents. A primary example of such technology is the electronic archive CATS providing open and broad access to

[73] Based on 2003 interviews.

a pool of unclassified information to all SIGNET-D users. Given the existing sense of vulnerability which emerged not only due to external threats of cyber-attacks on the system of the Department, but also due to the growing concern with the broad accessibility of the CATS system by anyone in the Department, many foreign service officers had simply chosen to continue enacting the need to know principle on their own. As a senior foreign service officer pointed out to me:

> The need to know principle has not been abandoned, not even temporarily, it has been there all along. The effect that the broad and open access to documents in CATS had, was that people simply stopped using CATS and rather stored info on their own personal drives.

In the words of another senior officer:

> The IT revolution has brought about an information overload and people simply react by not posting information on the intranet, by hoarding information in their small personal archives. Is this a backlash? No. Let's just call it a reaction to the new situation.

Hence, while there may have been a "watering down of classification awareness" in the name of faster and more efficient communication, there was no watering down of the general norm of secrecy and the culture of need-to-know that characterizes the foreign service. When the principles of openness and broad access that characterized the CATS system up until 2003 collided with the norm of secrecy and the need to know principle, it was the latter two that prevailed in spite of the fact that the technological infrastructure and procedures attached to it actually promoted openness and broad access. Foreign service officers, socialized into the diplomatic culture, chose to enact the available technology in their own way—not to publish documents even if they often were unclassified and wait to be asked by those in need of information and eventually provide it. Hence the practice of "fishing around for documents" discussed earlier, the effects of which clearly did not help the overall flexibility and effectiveness of information management.[74] Given the institutional setting informed by the diplomatic and bureaucratic norms, the result of the implementation and use of CATS was

---

[74] Besides the diplomatic norms, the bureaucratic norms also had its effects upon the uses of CATS. As one of the informants observed, a number of officers did not want to place the budgets of their organizational units "out into the open" as this would have had negative effects on their standing in the internal bureaucratic turf-battles that occur annually around budgeting.

not an effectivization of intra-organizational information management by broad access and openness that the technology enabled, but rather a *paralysis of information management due to too much openness*.

A solution to the problems indicated above—decreasing classification awareness, the risks related to chain e-mails and not least the low level of usage of the electronic archive (CATS) due to its overly broad accessibility and openness—has been the introduction of the above mentioned electronic records management system *Info Bank*. As pointed out above, development and implementation of this project was not managed by the technical support personnel but by FSOs. Hence, the system that was developed and implemented eventually took more into consideration the professional norms and values associated with the foreign service, including secrecy concerns and the need-to-know principle.

Following a search, all matching documents in the database were shown, but not all may be accessible. In those cases, an officer in need of a particular document to which s/he did not have access could request access from the officer who filed that particular document. The extent to which the need-to-know principle would be applied was hence determined by each and every individual officer producing and filing a document.

The Department had not introduced any authoritative rules or directives prescribing who should be granted access to documents filed in InfoBank. What was developed and made widely available was a set of best practices such as for instance that access rights should normally be granted to officers in the same organizational unit and the supervisors. Indeed, such a practice was very much in accordance with the need-to-know principle as practiced on paper. The chief difference that Info Bank seemed to make in terms of access was that it provided a better overview as to what information was available throughout the organization and where in the organization it was located, while maintaining the need-to-know principle. This might be considered a form of *guarded openness*.

4.6 *IT-effects on Communication with the Public*

While public affairs were still "a poor step child of the [Canadian] diplomatic service" in the late 1980s, the situation began to change following the end of the Cold War (Page 1993).[75] The early 1990s saw the establishment of the so called *Corporate Outreach Program*, which was later followed by *public diplomacy* becoming one of the core eight business-lines around which planning, reporting and managerial accountability at DFAIT were organized (see Table 4 above).[76] The advent of the Internet has been a welcome opportunity for the Department to increase its public presence. The Departmental web-site went online as early as 1995 and became thus one of the first foreign ministry web-sites in the world.[77] The Internet and e-mail have had effects on DFAIT's communication with the public both in terms of the direction of communication and in terms of the units communicating with the public.

4.6.1 *Direction of communication*

DFAIT has been using the Internet as a platform for a broad range of activities aimed at facilitating communication with the Canadian public. Two IT-supported initiatives stand out as innovative attempts to involve the public in discussions of foreign policy—the so called Dialogue on Foreign Policy and the Canadian International Policy Web-Site.

---

[75] As Page (1993:93), a former foreign service officer who gave more than a thousand public speeches at various forums throughout Canada during his 16 years in the Department points out, "assignments to information work of any kind were to be avoided like the plague or a kiss of death for advancement in the foreign service. Those who did get sidelined into this work were, and still are, regarded by the majority of their colleagues as second rate Foreign Service Officers who could not make it in the real world of diplomacy." In this respect, the practices of the Department were in sharp contrast with the practices of the Department of National Defence and CIDA, where domestic promotional activities were among the factors considered positively when promoting employees and who introduced training schemes developing employees' public speaking abilities (ibid., pp. 93–96).

[76] For an analysis of DFAIT's efforts to coordinate Canada's public diplomacy see Bátora (2006a).

[77] In 2003 the DFAIT web-site comprised more than 180 subsidiary web-sites (160 in 2002) and 80 mission web-sites in English, French and 14 other languages. The main site contains 80.000 pages (75.000 pages in 2002). Between 2001 and 2002, the number of visitors had increased by 110% and the Departmental web-site received over 1 million visitors per month in early 2002. This number had increased to 1.5 million visits by April 2003.The cluster of 180 web-sites linked to the Departmental web-sites received 12,9 million visitors in 2002. All numbers from *Performance Report 2001–2002*, Ottawa: DFAIT, p. 113; *Performance Report 2002–2003*, Ottawa: DFAIT, pp. 21, 106.

4.6.1.1 *The foreign policy dialogue: old wine in new bottles?*

The tradition of dialogues with Canadians on foreign policy priorities was initiated during the Trudeau administration in the 1970s. Following up on this tradition, the 1994 National Forum on Canada's International Relations sponsored by the ministers of foreign affairs, trade and national defense involved discussions with the Canadian public, which produced the strategic paper *Canada in the World—Canadian Foreign Policy Review* published in 1995. According to its Preface,

> [e]nsuring Canada's success as a society in a changing world must be a shared enterprise. The future of each one of us depends on it. That is why the Government is pledged to an open foreign policy process.[78]

Building on this legacy, DFAIT was actively engaging a broad array of domestic actors in institutionalized forms of consultation and cooperation with the aim of strengthening the legitimacy of foreign policy processes. Technological developments have greatly enhanced the ability of the Department to reach out and engage domestic constituencies on a continuous basis. The departmental web-site has been operational since July 1995 and has since then been one of the trend-setters among foreign ministries world-wide. Various tools have been employed to facilitate communication with the Canadian public.

One of the key initiatives was the so called *Dialogue on Foreign Policy* conducted from January to May 2003. DFAIT published a Dialogue Paper outlining Canadian foreign policy priorities and formulated 12 questions for public discussion concerning these priorities and values that Canada should stand for internationally.[79] An *eDialogue* web-site (www.foreign-policy-dialogue.ca—no longer operational) was launched, where citizens could download relevant documents and post their opinions and ideas on interactive message boards organized by themes. Citizens could also e-mail in their suggestions and propositions (and could of course also use ordinary mail for sending in their contributions). In addition to the electronic discussions there was a series of town-hall meetings across the country attended by the minister of foreign affairs and other governmental ministers, as well as 19 expert roundtables on

---

[78] *Canada in the World—Canadian Foreign Policy Review*, 1995 (http://www.dfait-maeci.gc.ca/foreign_policy/cnd-world/menu-en.asp)

[79] Processes such as these generating legitimacy for foreign policy approaches are of key importance for countries such as Canada, who aspire to take on the role of model-states or *gidsland* (mentor-state) promoting good governance. As Fossum (2006:785) argues, "a state's decision to behave like a gidsland has obvious implications for the substance as well as the process of policy formulation".

issues related to the Dialogue. The Dialogue web-site featured material from these expert roundtables including netcasts, as well as weekly summaries of contributions. The web-site registered about 62.500 visits, the Dialogue Paper was downloaded about 28.000 times and nearly 2000 people registered as participants on the online web-forum, which upon completion featured approximately 3.500 replies to dialogue questions.[80]

To foster a third party independent analysis and evaluation of the submissions in the electronic discussion forums, DFAIT engaged a non-state body called *byDesign—e-Lab*.[81] DFAIT also ran a parallel analysis of the submissions. The result is the ministerial report *Dialogue on Foreign Policy: Report to Canadians* presented in June 2003.[82]

The electronic discussions and town-hall meetings represent an unprecedented way of directly involving citizens, civil society organizations, businesses and sub-state actors such as provinces and municipalities in forming basic strategic guidelines of foreign policy. As the minister of foreign affairs writes in the introduction to the Report:

> ...the widespread engagement in town halls, on the Web site and in written submissions reaffirmed for me how strongly Canadians believe that direct citizen involvement must remain central to sound government, in the making of our country's foreign policy as well as in the reform and renewal of multilateral forms of governance. The advice summarized in this report will be vital to the work of policy development that will proceed in the months ahead. At a critical time in global affairs, your contributions will help guide our foreign policy and strengthen Canada's voice abroad.

The process of the Foreign Policy Dialogue as such provides an example of two-way communication with the public. As one of the senior DFAIT officers pointed out to me:

> The new thing about [the Dialogue] was the fact that the traditional mode of communication with the public—the press release mode—was abandoned in this case and one has tried to engage in an actual discussion with the public.

However, the actual level of the civil society actors' influence on the foreign policy strategies pursued in the end is ambiguous and the impact

---

80 http://www.dfait-maeci.gc.ca/cip-pic/participate/dialoguereport-en.asp

81 *byDesign—e-lab* was created by Liss Jeffrey, director of the McLuhan Program Research Network at the University of Toronto, see www.bydesign-elab.net, www.mcluhan.utoronto.ca/lissjeffrey.htm

82 http://www.foreign-policy-dialogue.ca/pdf/FinalReport.pdf

of the Dialogue on the foreign policy making process should therefore not be overestimated. Evaluation of the inputs and the process that led to the production of the above mentioned Final Report were far from transparent and according to interview information, the text of the Report was drafted more or less by one person on the Parliament Hill. The Report was presented to the Government and the public on a late afternoon before a long-weekend and received therefore relatively meager attention. What is more, even internally in the Department, the Report was received with some hesitation. As a senior DFAIT officer observes:

> The notion of the Dialogue on Foreign Policy has unfortunately not been credited enough attention by the senior management and by the Department in general. Sometimes this has to do with personalities, but the most important obstacle is the culture at DFAIT—diplomats simply feel discomfort when having to address the public. [...] There is a buzz-word very often used: 'deliberations inform our policy making'. This is a rather ambiguous word that simply does not disclose to what extent the deliberations have real impact—they can just be noted and passed by with no regard to them, but they may also have more substantial impact.

This seems to be consonant with the view of Hay (2000:4), who suggested that

> [t]he department's ministers, deputy ministers and assistant deputy ministers often assert their commitment to consultation, transparency and accountability. Yet there is no strategic, overall framework to activate and coordinate DFAIT's consultation performance. Neither (except in the longer hours worked by the relatively few officers personally charged with consultation duties) is there evidence of significant new resources directed to consultation. To declare the high importance of consulting Canadians, and then not to practise the principle, does look like a kind of structural hypocrisy.

As a follow-up on the Dialogue, the Strategic Branch at DFAIT had launched a series of eDiscussions on Canada's foreign policy priorities.

4.6.1.2 *Policy eDiscussions: public consultation and domestic network building*[83]

For several decades, the Policy Planning staff of DFAIT has been an active contact point between the Canadian public and the Department.

---

[83] This section provides a brief review of the eDiscussions process and practice at DFAIT. For a more detailed analysis of the eDiscussions see Bátora (2006b).

On a regular basis, academics, experts from other government departments or other specialists have been seconded on the policy planning staff to conduct analyses and bring in innovative ideas (Balloch and Angell 1992:458). As these insiders further observe, the Policy Planning unit had been serving

> as a window through which new ideas penetrate into "Fortress Pearson," as headquarters is sometimes called by some in government and the academic community, who believe that new ideas percolate out of the department more effectively and with greater frequency than they are allowed to filter in. [...T]he planning staff attempts to add oxygen to policy discussions within the department (ibid.)

The latest initiative in this spirit has been the initiation of the so called *eDiscussions* on the Canadian International Policy web-site maintained by the Strategic Branch of the Department (in charge of policy planning). It followed up upon the experience with web-based public consultation conducted as part of the Foreign Policy Dialogue. As stated in a November 2003 report by the Strategic Policy Branch to the Assistant Deputy Minister for Strategic Policy,

> [t]he Foreign Policy Dialogue web site was a signal to Canadians that DFAIT is seeking input from the general public into what is generally perceived as an exclusive and secretive process. Using Internet technology in consultation can seem to be radical to a department that is traditionally cautious. Therefore it is necessary to build a comfort level with online consultation by developing a well thought through process that requires minimum additional resources while opening up new doors for Canadians to have their voice heard by the policy makers.[84]

The tool for enabling the proposed kind of online consultation was to be the Canadian International Policy web-site featuring an eDiscussions-section. In the developmental stage of the web-site in late 2003, the goal was to establish it as "*the* focal point for all interested in Canada's international policy" (italics and bold letters in the original). Its aims were to be three-fold:

- *Proactively and Reactively frame* and *interpret* DFAIT's foreign policy for *public consumption*,
- Provide the *opportunity* and *context* for *policy discussion* and

---

[84] *Policy Planning Web Site Strategic Report*, internal memo, DFAIT, November 19, 2003, p. 6.

- Become the principal *resources* for finding information on *current* and *past Canadian international policy*.[85]

Besides supplying the citizens with an ultimate set of resources related to Canadian foreign policy and thereby giving them the necessary background for a qualified debate, the suggestion was for the web-site to feature issue-related topical discussion forums, where the participants would know that a one-paragraph synopsis of their contribution would be added to the minister's briefing notes.

The site was made operational in early 2004.[86] It did feature a well organized and searchable archive of policy statements, position papers and speeches but did not yet feature any of the proposed functions for public consultation or engagement. As it turned out, this was to be implemented in Phase Two of rolling out the web-site as of early November 2004. The new site includes the *Feature Issue* section, where topical issues for public discussion were posted by DFAIT. The public could then post their opinions on the web-site for a particular period of time (usually about 4 weeks).[87] Upon completion of an eDiscussion period, citizens' online contributions were summarized by Research staff of the Department. The summaries are then circulated to the so-called "Grad Student Committee" for review and comment, and returned back to DFAIT. Policy Research then completed the summary and circulated it to appropriate parties internally at the Department. Relevant officers then drafted an official reply to the summary and the draft would then be posted on the Policy Planning web-site. A new discussion would then be opened and old ones were archived and left accessible on the web-site for future reference.[88]

---

[85] Ibid., p. 3 (bold letters in the original).

[86] http://www.dfait-maeci.gc.ca/cip-pic/menu-en.asp

[87] What was to be posted as a Feature Issue was determined by an executive committee including the Assistant Deputy Minister for Strategic Policy, the Director General for Policy Planning, the Director General for Communications for Foreign Affairs and International Trade, the Director for Policy Research and Outreach, an invited representative from the Foreign Affairs branch that co-ordinates the specific content, the eCommunications Strategist of the Policy Planning bureau. Topics of Featured Issue between November 2004 and June 2005 included: "Renewing Multilateral Institutions", "Security" and "Showcasing Canadian Talent and Know-How Abroad".

[88] For a description of the whole process see http://www.dfait-maeci.gc.ca/cip-pic/current_discussions/input-en.asp

To create awareness about the site and get citizens involved from the start, three months prior to launching the site (in July 2004), Policy Planning Branch had posted a web-site proposal on the Internet and started to contact Canadian universities, NGOs and provincial authorities. The aim was to find interested individuals who could serve as "contact points" for the future policy discussion process. At the universities the persons playing the role of the Departmental "contact point" would be responsible for providing feedback on the web-site content to the Department, alerting students and faculty on topical issues posted on the Canadian International Policy web-site, and for communicating with other "contact points" across Canada. The process of disseminating the call for volunteer contact persons was initiated through a list serve message sent through POLCAN—"a list serve for political scientists"—run by the Canadian Political Science Association. Through this process, a network of individual volunteers working as DFAIT's contact points throughout Canada was created.

It is interesting to note DFAIT's effort at supporting the outreach web-based activities through human input "on the ground". Although the outreach activity and even the initial contact with the "contact persons" and organizations throughout the country was in electronic form referring to web-sites for further information, DFAIT's efforts indicate that reliance on technology alone does *not* guarantee the desired level of citizens' involvement. Again, the potential of web-technology needs to be harnessed by dedicated individuals, who through their personal contacts in their professional community would enable DFAIT to create greater awareness about its initiatives and ensure a higher level of responsiveness than would be the case without this human element in supporting the web-supported outreach activity. Clearly, this indicates that DFAIT's effort at citizen consultation and eventually engagement is not mere 'window dressing', but a serious attempt at creating a fairly stable structure of contacts throughout the Canadian society that would facilitate DFAIT's interactions with the domestic constituency.

In the period November 2004–November 2005, the International Policy web-site had 260.000 visitors, more than 2 million page views were recorded and the average time of a visit was 17 minutes. The International Policy Statement of the Canadian government was downloaded 134.000 times (123.000 in English, 11.000 in French) and its specific component on Diplomacy was downloaded 47.000 times (40.700 in English, 6.300 in French). In addition to accessing these resources, there were in February 2006, 2800 subscribers to the Canadian International Policy monthly eNewsletter, which represented a substantial increase from 1700 subscribers in February 2005.[89] Furthermore, engagement of the younger generations is also facilitated through the introduction of *pod-casts* in 2006, which allows the public to download material for listening or viewing on i-Pods virtually anywhere. Asides from the technical advancements, the attractiveness of the site was also enhanced by the dedicated approach of DFAIT staff in charge of eDiscussions. For instance, in the wake of the deployment of 2300 Canadian troops in the Afghan province of Kandahar in 2005, DFAIT's e-communications strategist (Mark McLauglin) flew to Afghanistan to film footage and gather other resources on the ground, which were then posted on the eDiscussions site as background resources. Moreover, representatives of other federal departments including defence and aid were also involved in the developments of the site contents and hence contributed to framing the public discussion (Garson 2007:215).

Asides from these positive strides, one also needs to note that there may be variations as to what issues will be put up for public debate. Strategies of bilateral relations with other countries or regions have so far not been put up for discussion. This may have to do with the closed nature of the foreign policy planning process in the geographic units, featuring mostly FSOs and informally having the highest status in the Department. As several of my informants pointed out, policy related suggestions in a discussion forum featured on the official site of the Department might confuse foreign governments as to whether what is posted there is the official position of Canada or not and unpredictable situations might arise. Hence, as one of my informants observed, reactions of FSOs to the idea of public electronic forums for policy planning was very often the following:

> What? Are we gonna get reactions on what we're thinking, and not on what we're doing?

The processes of transforming public inputs into the actual development of foreign policy strategies also play a role which limits the possible impacts of public consultations. As another informant pointed out:

[89] The numbers presented here are based on Mark McLauglin: "Canadian International Policy Web-Site: Public eDiplomacy", presentation, Policy Research Division, FAC, February 2006.

> Policy planning may be discussed in open forums, but when actual policy planning takes place, the doors close and everything is decided among two-three people in the room.

There seems to be variation in the potential impacts of public consultation in relation to various foreign policy sectors. As indicated by the interviews I conducted in 2003, geographical bureaus (the sites of the most traditional bilateral diplomacy) are far less inclined to use public inputs when developing strategies towards particular countries or regions than units dealing with the 'new' agenda such as human security, poverty reduction, migration, developmental issues etc. One of the reasons for this may be, as one of the respondents put it, that "geographic units in their development of strategies simply do not see a reason to take the opinion of the public into account. There are no incentives." While on the other hand, units dealing with the new agenda, see public consultation and eventually engagement as a strategic resource that can be harnessed and used as an instrument to achieve foreign policy goals.

In sum, there are several innovative aspects that IT brought into DFAIT's public consultation processes. Firstly, the amount of governmental documents and policy information that were made available to every participant (in practice any individual citizen) on a 24/7 basis had radically increased, which significantly improves chances for ordinary citizens to get involved in policy consultation as qualified participants. Secondly, citizens' inputs to the online forums could be reviewed by other participants at any moment in the consultation process and hence the consultation was not only a process between DFAIT and the citizens, but in fact a process of a general public discussion where citizens can react on each others' inputs. Thirdly, the time frame for conducting such large scale consultations with as many inputs as the Dialogue web-site had received, would be far longer if conducted without IT. Fourth, although, as indicated in the above, questions could be raised about the process that led to the production of the Final Report, the Fact is that the Final Report constitutes a strategic document guiding the official Canadian foreign policy in the coming years. IT-enabled public consultation hence did have some form of direct implications for the formation of Canada's strategic foreign policy priorities.

Despite these ambiguities in the actual impacts of IT-supported public consultations on the process of DFAIT's policy planning and decision-making, it is clear, that the Foreign Policy Dialogue and even more so the eDiscussions represent a shift towards *multi-channel ex-ante communication* between DFAIT (and/or its parts) and the Canadian public where the aim is not merely consultation but also engagement of the public.

4.6.2 *Level of centralization of public communication*

Since 1983, Canada has had its *Access to Information Act*, which specifies the conditions subject to which Canadians have a right to access information held in various forms by federal departments. To foster the contact with the Canadian public, the Department has set up an *Enquiries Center*, which was handling most requests for information submitted by the domestic constituency by various means of communication. On the DFAIT web-site, the Enquiries Center has a system, where the public can submit questions pertaining to particular topics and the system will automatically forward these requests to appropriate officers. As a further point of contact with the public, the main *Departmental library* served as a point of information provision to academics, journalists or other interested individuals.[90] Academics can also make use of the *Informal Access Program* administered by the Historical Section at DFAIT, which provides them access to documents outside the formal procedures of the access to information legislation. Finally, the *Media Relations Office* featuring about 10 spokespersons (each specializing on particular foreign policy portfolio) provided information to the media. While DFAIT does follow the provisions of the Access to Information Act, it does so in specific ways that have to do with its status as a foreign ministry. As a 2005 internal report points out,

> [d]ue to the international role of FAC and ITCan, the Departments face special challenges in the administration of the ATIP [Access to Information and Privacy Protection] legislation. The interests of other

[90] While conducting fieldwork for the current study, the library proved a valuable source of documents and information for the author. As a standard service, I was allowed unrestricted access to photocopying material available on the shelves and the staff were service-minded and effective in helping me find relevant documentation. Overall, the main library at Headquarters has about 40.000 walk-in visitors per year. This includes both internal clients (e.g. foreign service officers) and external clients such as journalists, academics or others. In addition to the main library at Headquarters, there are libraries servicing foreign publics at missions abroad in Washington, Mexico City, Paris, London and Tokyo. See *Audit of Departmental Library Services*, Audit Division, FAC and ITCan, October 2004.

> states and international organizations as well as Canada s bilateral relations would be seriously affected were sensitive information released inappropriately.[91]

Hence, as this report further observes, access to information requests need to be handled with special care and sometimes involve consultations with missions abroad and through them consultations with foreign governments, which in some cases could cause delays in DFAIT response times. Indeed, in the annual report for the fiscal year 2003/04 prepared by the Information and Privacy Commissioners of the Government of Canada including 'report cards' on the individual departments' compliance with the standard response times to access to information requests, DFAIT got a 'D' rating (regressing from 'B' in 2002).[92] The Access to Information Office at headquarters had 11 full time staff in 2003. There was a steady increase in the access to information requests from 386 in 1998 to 529 in 2002 (ibid.).

Besides these formal organizational procedures and structures related to the access to information legislation, a growing volume of public information provided by the Department was reaching the citizens through issue-related or client-group focused web-sites maintained by individual bureaus or programs of DFAIT.[93] One of the pioneers of such activities was the section in charge of relations with the European countries, which has developed its own profile and logo for the web-sites, all under a common name Canada*Europa* and under the banner of the Government of Canada and not under the banner of DFAIT.[94] The Section running the web-site encouraged also Canadian embassies in EU member states and other countries located in geographic Europe and adjacent regions to adopt their Internet graphical profile and logo. As one of my informants pointed out to me in 2003, one of the instruments for pressuring the embassies in Europe to adopt the Canada*Europa* look

---

[91] *Review of the Access to Information—Privacy Protection Function at Foreign Affairs Canada—International Trade Canada.* FAC—ITCan, Office of the Inspector General, Evaluation Division, March 2005, p. 4.

[92] *Review of the Access to Information*, p. 5.

[93] This development was related to a more general decentralization of the Departmental public communication in the early 1990s, due to which there were in 2000 about 30 units at DFAIT producing public material in parallel featuring "numerous design elements, variable editorial and printing quality and an inconsistent overall image. See *Evaluation of DFAIT Publishing*, Inspector General's Office, DFAIT, October 2000, p. 9.

[94] http://www.dfait-maeci.gc.ca/canadaeuropa/menu-en.asp

at their web-sites, was that only those embassy web-sites which adopt the new look would be included in the list of embassies featured on the opening page of Canada*Europa*.[95] In this way, a sub-unit identity has been developing in addition to the overall corporate identity represented by the common look and feel of the Department and at a more general level of the Government as a whole. Such a sub-unit identity (or what DFAIT called *second level branding*)[96] is of significance, because the Canada*Europa* web-sites have been generating over 20% traffic (or 250.000 visitors) to the Departmental site.[97] In general, this leads to multiplication of channels of DFAIT's public communication, where the above mentioned eDiscussions web-site maintained by the Strategic Branch is another example. One might hence argue that effects of IT have led to *increasing decentralization of DFAITs public communication.*

4.7 *Conclusion*

In sum, while hierarchy as a principle enabling political accountability remained in place at DFAIT, implementation and use of IT enabled increasing informational exchange across formal bureaucratic boundaries of organizational units and across hierarchical line relations internally. Broad availability of e-mail and Internet access and of electronic sharing systems supported by the InfoBank software package enabled work in virtual teams and integration of informational resources between units at headquarters and between embassies and particular units at headquarters. Project task-forces including external actors had been increasingly common. These findings match closest with the *transformation scenario* in the above developed analytical framework.

Secrecy remains a central norm informing the work of DFAIT. Increasing reliance on IT-tools led DFAIT to move away from risk

[95] For instance the Canadian embassies to Belgium, Czech and Slovak Republics, the Baltic states (in Latvia), Poland, Portugal, Spain, Switzerland and a number of other countries in Europe are not featured among the embassies that Canada*Europa* web-site refers to on its opening page. In August 2004, neither of these embassies featured the standard Canada*Europa* look. On the other hand, the Canadian embassy in Kazakhstan is on the list given the fact that it does feature the common look.

[96] http://lbp.dfait-maeci.gc.ca/europe/publica/publica.asp#p8. For basic guidelines on the look and feel of a mission web-site complying with the CanadaEuropa standards see http://lbp.dfait-maeci.gc.ca/europe/publica/look.asp.

[97] *Performance Report 2002–2003*, Ottawa: DFAIT, p. 106.

avoidance and towards *risk management* in its approach to electronic risks. The need-to-know principle was reinvigorated in the electronic systems for management of files and the policy of internal access to information matched the notion of *guarded openness*. In sum, the findings on the magnitude of IT-effects on secrecy at DFAIT match the *renewal scenario*.

Finally, while press-releases and one-way communication continued to characterize parts of DFAIT's public communication, innovative use of web-sites and other IT such as e-mail had enabled DFAIT to conduct broad *ex-ante* public consultations and engagement regarding foreign policy priorities and strategies. The practice of establishing agenda-specific web-sites by units within DFAIT has also led to emergence of *multiple channels of public communication*. These findings match the *transformation scenario*.

Following this discussion of one of the vanguard foreign ministries in the realm of IT-supported organizational reform, we can now turn to one of the latecomers—the Norwegian MFA.

CHAPTER FIVE

ROYAL NORWEGIAN MINISTRY OF FOREIGN AFFAIRS

Then we found out in disbelief that the information we provided was scribbled on pieces of paper, there were pieces of paper disorderly lying all around the table. One stack was held in place by a cup of coffee. Somebody took a sip of the coffee and the papers flew around in all directions! All of us helped to collect the pieces of paper so that information would not be lost. [...] I really thought that in 2004, one of the world's richest countries would come up with a portable PC, phone-service and Internet/mail. But this did not happen.[1]

5.1 *Introduction*

In recent decades, Norway has become one of the world's leading information societies. As a 2004 OECD report points out, "Norwegians are rapid adopters of new technology in general and new consumer electronics have penetrated rapidly. [...] In 2003, two out of three households had access to a PC at home and every other household had access to the Internet from home. 1.1 million individuals (one third) submitted and signed their tax return forms electronically in 2003 (Skatteetaten, 2004), and 1.9 million customers had Internet bank accounts in 2002 (Norges Bank, 2003)—near the top internationally. Three in five Internet users purchased goods over the Internet in 2003. [...] Thus the population is well versed in using ICT [...]."[2] E-governance initiatives have been a top priority for virtually all Norwegian governments in recent years and agencies of the Norwegian public sector have been leading internationally in adopting and using technologies.[3] It comes

[1] A Norwegian tourist describing the interaction with the Norwegian embassy staff in Phuket, Thailand following the tsunamis in December 2004. From the report *26.12. Rapport fra evalueringsutvalget for flodbølgekatastrofen i Sør-Asia* [26.12. Report by the Evaluation Committee for the Tsunami-Catastrophe in South-Asia] Oslo: Statens Forvaltningstjeneste, April 2005, pp. 22–23 (author's translation).

[2] *ICT Diffusion to Business: Peer review. Country report: Norway*, OECD, 2004, pp. 11–12 (online at http://www.oecd.org/dataoecd/32/7/34228124.pdf).

[3] See for instance the 2002 strategic governmental report *eNorway 2005* at http://www.regjeringen.no/upload/kilde/nhd/rap/2004/0029/ddd/pdfv/198680-enorway_2005.pdf.

hence as somewhat of a surprise that the Norwegian Ministry of Foreign Affairs (N-MFA), had been a severe laggard in comparison with other agencies of the Norwegian government and even in comparison with other foreign ministries in implementing and using IT in support of its operations.[4] In the context of the current study, it is interesting to explore the effects of IT on a foreign ministry characterized by a strong *esprit de corps*, strict adherence to traditional diplomatic norms, structures, routines, and by traditional scepticism to organizational modernization (Ibsen 1911, Omang 1954, Colding 1978, East 1981, Christensen and Egeberg 1989, Blichner and Sangolt 1993, Christensen 1996, Neumann 1998, Neumann and Leira 2005).

This chapter first outlines the institutional characteristics of N-MFA. The IT infrastructure of the Ministry is then described. This is followed by a discussion of the effects of IT on the Ministry by focusing on the three core organizing principles of diplomacy. Conclusions follow.

5.2 *Institutional Background and Characteristics of N-MFA*

Although a fully fledged Norwegian foreign ministry was established first on June 7, 1905, that is on the very same day as the dissolution of the Swedish-Norwegian Union was announced, Norway did conduct separate quasi-diplomatic activities prior to that date. This was facilitated by the fact that Norway did have its own separate parliament, government, civil service and military forces; all anchored in Norway's 1814 constitution. Norway's foreign affairs was one of the few areas in which the Swedish king had retained the ultimate authority (Norman 1982:392). Nevertheless, as of 1858, there had been an office of consular and commercial affairs within the Norwegian Ministry of Interior in Christiania (Oslo) handling primarily foreign commercial activities of Norway. In 1899, the Department of Commercial, Consular and Foreign Business was set up within the same ministry. Then in 1901, the Ministry of Interior itself was re-named to Ministry of Foreign Affairs, Commerce, Shipping and Industry (ibid., p. 393).

---

[4] See the strategic report *The Electronic Foreign Service*, Oslo: Ministry of Foreign Affairs, June 25, 1999 (online at http://www.regjeringen.no/nb/dep/ud/dok/rapporter_planer/rapporter/2000/The-Electronic-Foreign-Service.html?id=420085). See also Bátora and Neumann (2002).

Several Norwegians had served in the foreign ministry of the Union in Stockholm as well as at Swedish/Norwegian embassies. Thor von Ditten, for instance, was the permanent under-secretary of state (the highest administrative office) at the foreign ministry in Stockholm in the period 1901–1903. Given his experience with running the administration of the Swedish foreign ministry, von Ditten was instrumental in setting up the administrative capacities of the Norwegian foreign ministry following the dissolution of the Union in 1905. In the capacity of a special advisor to foreign minister Løvland, von Ditten suggested a two-pillar structure including a Department in charge of diplomatic and political affairs and a Department in charge of consular and commercial affairs, with an Archive unit placed separately between them (ibid.). The new organization carried significant imprints and organizational traditions from the Swedish foreign ministry, which prompted the radical liberal Wollert Konow to call it "a modest pocket version" of the Swedish model of foreign affairs administration (cf Omang 1954:26). The new Ministry moved into offices in Victoria Terasse in central Oslo, where it still resides.[5]

5.2.1 *Organizational structure*

Despite the growing number of actors involved in foreign policy processes and the shifting locus of foreign policy making resulting from that, the N-MFA retains its position as a central coordinator of Norway's foreign policy. This is related to the fact that the Norwegian political-administrative context, just as that of other Scandinavian countries, is relatively small and homogeneous which results in fairly centralized structures of foreign policy coordination (Sundelius 1984). In the Norwegian practice this means that while other government ministries are routinely involved in the administration of the international agenda related to their sector-specific expertise, the N-MFA retains the role of the top coordinator in situations where there is disagreement between the sectoral interests and the interests of the state (Knudsen 1997:80–81).[6]

---

[5] *Utenriksdepartementet historie* [History of the Foreign Ministry] available at http://www.regjeringen.no/nb/dep/ud/dok/rapporter_planer/rapporter/2000/Utenriksdepartementets-historie.html?id=105584; for the most comprehensive organizational history of the N-MFA available to date see Neumann and Leira (2005).

[6] As an example, Knutsen (1997:80) mentions a situation in 1995, when the Norwegian defense minister announced his intention to withdraw Norwegian peacekeeping troops from Lebanon. The foreign minister disagreed arguing that the troops are

Reflecting this principle, structural changes in the Norwegian government in response to increasing internationalization of the agenda administered by line ministries have hence not led to decentralization of foreign affairs administration, but on the contrary to further centralization of administrative functions at the N-MFA and to maintenance of the N-MFA's coordinator role (East 1981, Chrisensen 1996). This led to a series of administrative mergers. In 1988–90, first the Ministry of Trade and later the Ministry of Development Cooperation were merged with N-MFA. The rationale behind the amalgamations was to improve coordination between the agencies of the central government most frequently involved in international policy activities. The second phase was in 1997, when parts of N-MFA (a department including several sections in charge of trade policy and export promotion) were assigned to the newly reformed Ministry of Trade and Industry. The most recent set of organizational changes were related to the January 1, 2004 founding of *Innovation Norway* and its integration under N-MFA's jurisdiction,[7] and to the integration into the N-MFA of NORAD (Norwegian Agency for Development Cooperation) in February 2004.

The N-MFA has a mixed organizational structure, including geographic bilateral units and functional departments (see Appendix 5). At the time of writing, the N-MFA was led by two ministers—a Minister of Foreign Affairs and a Minister of International Development. Each of them was responsible for a different agenda, but in decision-making the Minister of Foreign Affairs was the highest authority according to the *primus inter pares* principle.

As a study observed in the mid-1990s, due to the administrative mergers and growing complexity of the foreign policy agenda, the N-MFA was much larger, more heterogeneous, with a larger workload and heavier media pressures than in the 1970s (Christensen 1996). This, though, had little effect upon the formalistic bureaucratic culture

necessary as a diplomatic tool in ongoing peace negotiations in a regional perspective, and eventually had his way.

[7] Innovation Norway is a state owned company replacing the Norwegian Tourist Board, the Norwegian Trade Council, the Norwegian Industrial and Regional Development Fund and the Government Consultative Office for Inventors. Innovation Norway's offices in more than 30 countries abroad were integrated into Norwegian embassies, who act as coordinators of all Norwegian activities in the countries in question. For more information see http://www.innovasjonnorge.no.

of the N-MFA (Knudsen 1997).[8] Knudsen (ibid, p. 76) relates this to the practice of rotation (*flytteplikten*), which reduces the ability of foreign service officers to create personal relations with colleagues in domestic ministries working on a specific issue area, a practice otherwise common between the administrators in domestic ministries of the Norwegian government. As he further suggests, the bureaucratic culture at N-MFA might also be strengthened by the generally formalistic style of diplomatic exchanges.

At the time of fieldwork in 2003, Norway had a network of 102 missions abroad. Of these, 84 were embassies, 8 permanent missions to international organizations, 9 consulates general, and 1 so called *interest-office* ('interessekontor'). In addition to these, there were 394 unpaid honorary consulates.[9]

5.2.2 *Personnel and organizational culture*

Historically, most of the members of the Norwegian foreign service had a law degree.[10] Following the 1922 reform of governmental administration, a law-degree was no longer required for entering into the foreign service. Instead, one has introduced the so called *aspirantprøve* that same year, which was a foreign service entry exam featuring a standardized set of questions on international law, political economy, history, diplomatic practice and foreign languages. The aim with this selection procedure, which is used until today (1–2 % of applicants are accepted every year), was to achieve a situation, where the foreign service would no longer be a "closed agency" ("lukket etat") and instead enable the entrance of qualified applicants irrespective of their educational background and social standing. Nevertheless, the prevalence of diplomats with a law degree was significant up until the late 1970s when candidates with social science background were increasingly admitted into the foreign service. While 28% of employees of N-MFA had a law degree in 1976

[8] Neumann's (2001) descriptions of speech-writing processes in the N-MFA observing strict bureaucratic procedures and face-to-face contacts exemplify this well.

[9] Based on *Utenriksdepartementets historie* [History of the Foreign Ministry] available at http://www.regjeringen.no/nb/dep/ud/dok/rapporter_planer/rapporter/2000/Utenriksdepartementets-historie.html?id=105584.

[10] All the senior officers who entered the Norwegian foreign service upon its founding in 1905 were lawyers. What is more, most of the junior officers, many of whom later came to play important roles in the foreign ministry, were also lawyers (Omang 1954:95–96).

and only 10% had social science background (mostly in political science), the numbers had practically shifted in the next ten years and lawyers represented only 11% and social scientists 26% of the N-MFA's workforce in 1986 (Christensen and Egeberg 1989:35).

Since the early years of N-MFA's operation, career ladders have been carefully observed. Foreign service officers traditionally embarked upon life-long careers, where the time periods between promotions were relatively long and seniority had been one of the main conditions for promotion. As Christensen and Egeberg (1989:42) found, compared with other ministries of the Norwegian government, the N-MFA had a disproportionately high number (40%) of officers with 20+ years of experience in the service in 1986. Stability of career paths has been increasingly under strain in recent decades due to growing complexity of the foreign policy agenda and expansion of the workload at the N-MFA. While the Ministry had maintained its strongly bureaucratic character with strict adherence to line authority relations and strict divisions between jurisdictions of organizational units, the growing number of special advisors and other experts taking on positions at the Ministry had been a factor undermining the traditional bureaucratic procedures in the day-to-day workings of the Ministry. As East (1981) had observed, in some cases this led to ambiguity about lines of authority. Further pressure has been coming from an ever increasing involvement of line ministries in administration of international agenda (Colding 1978, East 1981, Efjestad 1981, Blichner and Sangolt 1993).[11] In response to such suggestions, N-MFA has been insisting that there is a difference between 'foreign activity' and 'foreign policy', has been refusing to give up what they considered to be the key areas of foreign policy (e.g. issues of national security, human rights etc.), and has been claiming the prerogative to take over foreign policy mediation anytime they feel Norwegian national interests are at stake (Neumann 1998:80).

Due to its professional diplomatic specifics (peculiar procedures for selection and socialization, rotation of personnel, generalist expertise etc.), the N-MFA has been developing a sense of being different and unique in the context of the Norwegian government and the Norwegian civil service (Neumann 1998). Between rotational foreign service

[11] Some authors went as far as to suggest that the N-MFA should be dismantled, the coordination of foreign affairs agenda should be assigned to international departments of line ministries, and a small coordination unit at the prime-minister's office be tasked with the coordination of embassies (Moses and Knutsen 2002).

officers this has been fostering close collegial relations and an *esprit de corps*, which often disregard formal bureaucratic lines and instead build on a sense of belonging to a professional group. What is more, groups of foreign service officers who had passed through the candidate course at the same time often build informal 'cliques' within the system and individual officers can count on the support of other clique-members in informal information exchange irrespective of where they are located. Also, clique-members often compare their career development with that of other members, which also leads to an extra pressure upon the internal personnel policies for keeping the career ladders approximately alike for officers with the same number of years of experience in the service (Neumann 2001c). The close collegial relations between Norwegian diplomats and the relatively small size of the organization, where one after serving for some years usually does have substantial personal networks, enable informational exchange throughout the organization, which often disregard formal bureaucratic or diplomatic ranks and also formal communication procedures. As will be discussed in more detail below, informal relations foster greater willingness to share information electronically and even unsophisticated systems for electronic information management such as a simple e-mail system, provide great gains.

N-MFA strongly insists on preserving the exclusive right to man Norwegian missions abroad by its rotational employees ('flyttepliktige'—which in the literal translation means 'obliged to move') and also strictly keeps to the principle that someone becomes rotational (a professional diplomat) only and exclusively after passing N-MFA's candidate course (*aspirantkurs*). This is especially important in turf-battles with other Norwegian ministries, who attempt to assign their own representatives at Norwegian missions abroad (ibid., pp. 82–83). Given the fact that N-MFA does not have any stable and clearly defined domestic support constituencies such as the domestic ministries do have (e.g. fishermen community and the Ministry of Fisheries, farmers and the Ministry of Agriculture etc.) sets N-MFA apart from other Norwegian ministries (East 1981, see also Dahl Jacobsen 1963). A corollary to the the dominance of career diplomats in the organization is a strongly hierarchical culture and any career-minded officer avoids criticisms of the senior ranks. As a lead-article in the Norwegian daily Dagens Naeringsliv held on January 6, 2005,

> the Head of Administration [at the N-MFA] is said to surround himself only with yes-people. Alternative views or criticism is impossible without negative consequences. Noone dares to criticise the leadership. Senior ambassadors with many years of experience tremble when called in to the Head of Administration. Fear and anxiety, almost never any praise, those are the characteristics of the culture [of the N-MFA]....This is a corporate culture that has been abandoned in the society at large fifty or a hundred years ago.

In 2003, N-MFA had approximately 700 employees at the headquarters in Oslo. At Norwegian missions abroad, there were about 600 Norway-based staff including special envoys and about 800 local employees.

5.2.3 *Domestic environment*

Ever since its foundation, the N-MFA has had somewhat strained relations with the Norwegian public. When Norway declared its independence in 1905, large parts of the political spectrum did not want the country to have a foreign policy of its own so as to avoid being pulled into what was often referred to as the morally dubious power politics of the large states. The slogan of the day among the liberals and leftists was "We want no foreign policy!" Consequently, there was only meager support for the idea of having a foreign ministry. At best, Bjørnstjerne Bjørnson was arguing for a "foreign ministry without a foreign policy" (cf. Berg 1994:43–44). In the budget passed by the parliament (*Storting*) in 1906, severe cuts to the new foreign ministry's budget were introduced and many of the positions within the ministry had received a non-permanent status (Norman 1982:394–395). Pressures were also coming from the business community, who considered the consular and commercial service an asset, but the diplomatic service was seen as an inefficient organization whose ceremonies and luxurious lifestyle only produced expenses and hardly ever any returns.

Unwillingness to have a foreign policy was not the sole factor contributing to the popular skepticism towards the diplomats. The general public was strongly critical of the secretive character of diplomatic working methods and a number of central political figures sought to bring the foreign affairs administration under tight parliamentary control and scrutiny (Omang 1954:24). Moreover, as the diplomatic envoys in the early 20th Century still had to cover most of the expenses arising from their assignments abroad from personal resources, a diplomatic career involved a great deal of personal wealth, which led to all but democratic recruitment patterns. What is more, popular skepticism towards the diplomatic service was further strengthened by the fact that the Swedish-Norwegian diplomatic service prior to 1905 was for the most part manned by members of Swedish aristocratic families. As Neumann (1998) points out, until today the word diplomat carries a somewhat negative connotation and Norwegians prefer using the term 'UD-tjenestemann' (MFA-officer) when referring to Norwegian foreign service officers.

The perception of N-MFA by the Norwegian society was further problematized following the negative outcome of the 1972 referendum on Norway's EU membership (Neumann 1998:87–89). In the public debate prior to the referendum, the government used foreign service officers to inform about the positive aspects of EU-membership, and N-MFA thus came to be perceived by the general public as taking a strong pro-membership position. The following negative outcome of the referendum going hand in hand with youth radicalization based on anti-elitist attitudes of the late 1960s, put the Ministry in a negative light, because the diplomats were perceived as not representing the nation as a whole but rather their own elitist and political interests (ibid., p. 90). This brought about a situation where it became the daily bread of Norwegian diplomats to struggle with the high-level of suspicion and distrust that their Ministry had to face in its relations with the domestic population.

While the N-MFA is the lead agency in charge of foreign affairs on behalf of the Norwegian government, it is institutionally embedded in a political-administrative system which involves a host of other actors in the making of Norwegian foreign policy. Firstly, it is the parliament (*Storting*), which ratifies Norway's international treaties and also passes bills for the government executive, including the foreign affairs administration. Given that Norway has had a tradition of minority governments in recent decades, the Parliament and individual political parties have had a considerable role in influencing foreign policy decisions (Knudsen 1997:73–74). Secondly, interest groups and NGOs have been increasingly active in shaping international policy. Cooperation of the N-MFA with NGOs has taken various institutionalized and semi-institutionalized forms ranging from coordination meetings to financial support and staff exchanges (see Egeland, 1988, Tvedt 1997, Bucher-Johannessen 1999). Knudsen (ibid.) therefore suggests that while Norway's foreign policy traditionally used to be formed mostly by the N-MFA, the locus of foreign policy making has shifted in recent decades and foreign policy is now being formed in "a political space" between the government (N-MFA), the Parliament, interest groups and the media, where the role of the latter is shifting and unpredictable. Yet, as Neumann (2005) argues, the involvement of other ministries and NGOs in foreign policy cooperation had not led to any substantial changes of the organizational structure and culture at the N-MFA, which continues to be centred around the issues of national security as the core area of N-MFA's activity and responsibility while other policy issues remain on the fringes of the organization with difficult access to the senior ranks. This results in a policy issue hierarchy at the N-MFA, which perpetuates the dominant position of career diplomatic staff in the organization (ibid.).

5.3 *IT-infrastructure of N-MFA*

Introduction of IT at the N-MFA has been a gradual process of adaptation where strategic plans were adopted and frequently changed due to the rapidly changing character of the technological environment throughout the 1990s. The rationale behind the implementation of IT-supported reform processes within the organization was outlined in the 1999 Report *The Electronic Foreign Service* (further referred to as the EFS Report), produced by an internal committee.[12]

As observed by the authors of the EFS Report, there was a growing realization throughout the Norwegian Foreign Service that it seriously lagged behind other countries' foreign ministries in using IT, which could lead to strengthening of the perception in the Norwegian society of the N-MFA as increasingly irrelevant. Moreover, the authors of the EFS report also argued that immediate and extensive modernization of the N-MFA's IT-capabilities also "has to do with national interests" (p. 3). To reverse the trend and save the Ministry from irrelevance, the Report proposes a vision of the Norwegian foreign service working

---

[12] The full title of the document is *The Electronic Foreign Service. Proposal for a strategy for information and communications technology and information management at the Norwegian Ministry of Foreign Affairs 2000–2003.* Oslo: MFA, June 25, 1999 (available at http://www.regjeringen.no/nb/dep/ud/dok/rapporter_planer/rapporter/2000/The-Electronic-Foreign-Service.html?id=420085).

on a global scale as a "pro-active and knowledge-intensive network organization".[13]

With this objective in mind, a fairly wide modernization of servers and workstations was undertaken along with new hardware and software acquisitions.[14] An upgrade of the N-MFA's communications infrastructure was also needed. Before 2000, the electronic system of the Ministry used for daily production of documents—the so called *Black Network*—was not connected to the Internet for fear that classified information that it contained might be exposed to electronic attacks. Solutions for external e-mail communications were almost unavailable at headquarters and embassies, and personnel often used private e-mail addresses from their home PCs to communicate by e-mail with external contacts. Internet connections were provided from dedicated computers, the so called "Internet-kiosks" located in corridors of the Ministry—the so called *Blue Network*. The most significant step in terms of upgrading the electronic infrastructure was the *switching of networks* in 2000 following the findings and recommendations of the Electronic Foreign Service Report. Full internet and e-mail access from all workstations used in daily work by officers was provided through connecting the Black Network to the internet both at headquarters in Oslo and at missions abroad. At the time of writing, it carried unclassified internal traffic and provided access to the organizational intranet[15] (implemented as of June 2000 at headquarters and encompassing all missions abroad

[13] Quoted and translated from the Norwegian version of the report, which states "en proaktiv og kunnskapsintensiv nettverksorganisasjon" (p. 3). For unclear reasons, the official English translation features the phrase "an organization with a proactive and knowledge-intensive network", which evokes a slightly different meaning. Following consultations with one of the central authors of the EFS-Report about the meaning and rationale of the vision, I chose to stick to my own translation of the Norwegian text.

[14] Even though N-MFA's yearly IT-budgets have been increasing in the late 1990s (e.g. an increase from 32 to 47 million NOK per annum between 1997 and 1999), the Electronic Foreign Service Report observed that the amount spent was still only 50–70% of what other Nordic countries were using on their foreign ministries' IT-infrastructures.

[15] The Intranet was gradually developed and in 2003 featured internal newsletters such as *UD-info* and *News from Victoria Terrasse*, as well as other useful internal information including the Norwegian List of Diplomatic and Consular Representation Abroad, personnel policy guidelines, budget and accounting data, travel allowance rates, overtime rules, filing instructions, the staff handbook, the Oslo Diplomatic List, country reports etc. It also provided access to external online information sources that N-MFA was subscribing to such as Oxford Analytica, Reuters, Keesing's and other kinds of encyclopedias and reference works.

in late 2003). The switching of networks involved a process in which sensitive or classified information and documents were removed from the Black Network and migrated to the Blue Network, which was 'hardened' against electronic eavesdropping and other kinds of unauthorized electronic access by outsiders. The classified Blue Network is available through dedicated PCs placed in certain offices at headquarters and in certain missions abroad and only authorized personnel have access to this system.

Following the switching of networks, Norwegian missions abroad were informationally integrated into the systems used at headquarters, so that the user interfaces and access to some of the informational resources is the same irrespective of where an officer is located. In principle, this network infrastructure solution was similar to that earlier introduced by DFAIT (see previous chapter) and to the one later adopted by MFASR[16] (see next chapter).

At the time of the second phase of fieldwork at the N-MFA in 2003, the organization was in the third year of radical modernization of the infrastructure and IT-supported working procedures. The next section presents the findings about the effects of IT in N-MFA.

5.4 *IT-effects on Hierarchy*

Prior to the above mentioned "switching of networks" in 2000/2001, informational and communicational practices at the N-MFA and throughout the Norwegian foreign service were largely paper-based and PCs were mostly used as advanced typewriters, which maintained strongly bureaucratic routines. This led two insiders to describe the N-MFA as living in the "digital Stone age" (Solheim and Solem 1997). The high level of frustration among N-MFA staff had to do mainly with the limited availability of Internet-connections and e-mail communication in the Ministry and at embassies. Given the fact that the network used for daily production of documents contained also classified information and documents in electronic form, it was not connected to the Internet. Internet was available at the so called "Internet-kiosks" in

[16] It is of interest to note that the Norwegian telecommunications company Telenor had been the main provider of network infrastructure solutions to both N-MFA and MFASR, which hence leads to practically identical global technological platforms in both organizations.

the corridors or in especially dedicated computer rooms. Foreign service officers in need of Internet-access had to move to these dedicated workstations, which of course brought about severe inefficiencies, as officers often had to wait and line-up while others were surfing the net. Also, constantly having to remove all the working documents from one's desk to the Internet-kiosk and work there was seen as highly unsatisfactory. Facing these difficulties, many officers simply opted for not using the Internet at work and instead got Internet-access installed privately at home so that they would conduct the necessary web-searches privately in the evenings or early mornings and routinely beat their colleagues and superiors with updated information on developing foreign policy issues discussed at morning meetings.

The problems with using technology were not only related to the unavailability of Internet and e-mail connections, though. There was in fact an internal e-mail system installed on the production network (Blue Network) at headquarters, and about 400 employees were connected to the system from their own workstations. Despite its availability, the e-mail system was severely under-used. As Solheim and Solem (ibid.) reported, on a random day, Thursday, April 24, 1997 at 10.20 am, only 65 employees were logged on to the internal e-mail system. This might have had to do with the culture and habits at the N-MFA, where it was usual, that one would walk over and speak to colleagues in other divisions face-to-face when coordination about specific issues was necessary. Neumann (2001b:136) describes the face-to-face habits in a situation, where, as senior advisor on assignment in the Planning Unit, he was charged with writing a speech to be given by the Norwegian King upon his upcoming visit to Hong-Kong:

> I will find out who is in charge of Hong Kong in the Foreign Economic Policy department and send him or her an e-mail, I suggested. Well, says [my mentor], but it would be just as good if 'I walked over and talked to the desk officer in person'. The process should not be anonymous—face-to-face contact is important. So I set up a draft, print it, make a copy of the instruction-memo and of the memo from the Consulate in Hong Kong, check the phone book to find the desk officer in charge, call the officer and announce my arrival, and then I walk the 300 meters of corridors from the Northern Quarters of Victoria Terasse [...] over to the Southern Quarters, where the Foreign Economic Policy department is located. The officer and I have a talk; he tells me that he probably would find some background material that he would send over in the mail tomorrow. Two days later a brown internal envelope lies in my in-tray with half a page of text and a diskette. I sit down and write the speech,

> print it out, walk over to the secretariat and tell the secretary about the speech. 'You should put it up to him', she tells me. 'He' is the head of the division. He sits, as we all others do, physically located in the third floor, but his in-tray is highest up in the stack, and he is the head of the unit. Also in units, where the in-trays are lined up horizontally with the one belonging to the head of the unit being the one furthest to the left, one 'puts up' drafts for approval.[17]

Similarly at embassies, where prior to the "switching of networks" in 2000/1 there were usually only one or two PCs with Internet access, foreign service officers felt strong pressure from the environment where the use of IT for electronic communication was increasingly pervasive. As an officer at a Norwegian embassy in Latin America reported back to headquarters on February 15, 1999:

> The embassy perceives the current solution with one shared e-mail address as inefficient, because the incoming e-mails are for practical reasons usually printed out and given to the individual employees [on paper]. To answer, the staff have to either use a diskette or take the relevant papers to the multimedia PC and sit there and answer. We feel therefore that the increasing use of e-mail by persons and institutions in various contexts brings about more work, and that it is in fact easier to answer incoming e-mails by other means, for example by fax. [...] Increasing use of e-mail will therefore not necessarily lead to a decrease of the workload, unless every employee has an e-mail connection from their own workstation.[18]

This view was confirmed by a report from a Norwegian embassy in a major Asian country dated February 5, 1999:

> The embassy currently has a dial-up Internet/e-mail connection which works fine. It is at times difficult to get through in the middle of the day, though. Due to complicated procedures e-mail is not used in the communication with headquarters. Use of e-mail is also limited by the fact that there is only one PC with such a connection and it is located in the corridor. Fax and phone are still preferred by many [employees]. There is undoubtedly a need for a cost-effective communication system.[19]

Finally, a set of observations from a Norwegian embassy in another major Asian capital dated January 28, 1999, indicate the implications of the cumbersome e-mail and Internet connections for the very relevance of the embassy:

---

[17] My translation.
[18] My translation.
[19] My translation.

> What is important for the embassy is not the means but the speed and breadth of information [we have access to]. In this respect we feel we lag behind many of our colleagues [in other countries' embassies], which cannot simply be related to how advanced ours and theirs technological means are. Even though it no longer happens so often that reports about particular events reach us several weeks after the fact, it still takes at least a week before a report sent from a [Norwegian] mission [in another country] reaches us by fax. For example, the forwarded reports we received by fax today, January 28, are dated January 12, 16, 18 (2 reports), 20 and 21 (2 reports), which suggests that reports [reaching us] are on average 10 days old. In the era of Internet and CNN, the administrative routines of the N-MFA are so slow, that the use of fax can no longer guarantee topicality of the reports sent to us. [...] Given the fact that Norway, being a small country, is basically not considered highly important in most places, it can be advocated as crucial for a country like Norway, more than for many larger countries, to have well-informed and updated representatives at its embassies.[20]

As the above quotes indicate, paper-based bureaucratic procedures have been quite pervasive in the Norwegian foreign service. The so called "switching of networks" in 2000–01 had brought about a number of challenges to the bureaucratic line relations at N-MFA and to hierarchical relations between headquarters and missions abroad.

5.4.1 *Challenges to bureaucratic hierarchy*

It has to be said at the outset that N-MFA remains a bureaucracy with fairly clear divisions between jurisdictional areas and with strictly observed hierarchy of offices. Yet the radical expansion of both internal and external information flow that had followed the "switching of networks" in 2000 had brought about a set of challenges. First and foremost, access to e-mail and Internet came at a comparably late stage, when most actors with whom N-MFA had been communicating had been using e-mail as a standard form of communication for some years. The expected increase of electronic information exchange happened in the context of largely paper-based routines. As an internal report[21] of the N-MFA, delivered only days prior to the switching of networks, stated:

---

[20] My translation.

[21] *Rapport fra forprosjektet for elektronisk saksbehandling* [Report from the Pre-Project on Electronic Management of Documents], Oslo: Ministry of Foreign Affairs, May 31, 2000 (internal), pp. 8–9.

Today, the foreign service is basically not prepared for the development that will be brought about by the increased use of IT-services in the governmental administration. Information management and information-use is today characterized by:

1. *Paper-based archive*—Documents filed in the foreign service archive are only accessible on paper. There is no electronic database with official files. Officers hence do not have access to information from the archive beyond office-hours. Besides, it is time-consuming to find paper-documents in the archive. Most of the officers have therefore established their own personal handy-archives. This is unsatisfactory, not least for security reasons. Low accessibility of the archive also contributes to the fact that the archive is not complete. For example, the archive receives only a fraction of the memos that are produced. The fact that the archive is incomplete further strengthens the need for handy-archives in the offices [around the Ministry].
2. *Ineffective re-use of filed information*—An archive that only exists in paper form makes it difficult to re-use information. If a text from a filed document is to be re-used, it needs to be typed in anew, which is obviously severely unsatisfactory in today's IT-world. Re-use of electronically filed texts is today done based on files in shared databases or in personal computer files of individual officers. There are currently no shared guidelines for the saving of electronic documents and practices vary between divisions and between individuals. This leads to difficulties in finding and re-usage of documents.
3. *No automatization of routines*—No systematic assessment had been performed in relation to how routines in the organization would support the electronically supported management of documents. Those routines for electronic management of documents that are available are basically just a continuation of routines for the paper-based information flow. This means that the actual information flow is basically paper-based. A significant potential for rationalization therefore remains unused.
4. *No coherent solution in the field of information management*—The [foreign] service has a shared archive service in the Ministry, but also operates archives in every mission abroad. This lack of integration leads to a situation where the same document is registered and thereby also sorts under administrative authority in the Ministry and also in individual missions.

In an environment characterized by the prevalence of paper-based practices, a number of ambiguities arose in relation to the technological solutions that were made available. One of the problems was the status of e-mail communication. Similarly as in the case of DFAIT, there was often uncertainty as to whether e-mail messages were oral or written communication and consequently uncertainty as to what should and should not be archived. Another problem was related to ambiguities about archiving procedures to be applied in relation to electronically edited documents. It was not always clear whether one should file only the final versions of documents or also the work-in-progress versions featuring inputs and editing-comments from officers participating in the processes of producing a particular document.[22] As a result, there was an estimated 40% decrease in the information filed in archives between 2001 and 2002.[23] This situation has led one of the directors to argue that

> the years around the turn of the 21st Century will be considered the dark age by the historians of Norwegian foreign policy, because they will not be able to reconstruct the development of decisions. Much of the institutional memory which previously would have been kept orderly on paper in the archive, has evaporated in fragmented small archives on individual PCs within the Ministry. This is a problem also when it comes to accountability and responsibility for decisions, because no one really knows who has said what except for the people involved in the actual processes. If these people are [at the embassy] in Japan or dead, we have a problem.[24]

As this statement clearly indicates, effects of IT were seen as threatening bureaucratic norms of accountability and internal process-transparency enabling a ex post reconstruction of decision-making processes. In addition to the fragmentation of the institutional memory, IT also enabled multiple versions of documents to be produced and circulated. Hence, when working versions of documents were circulated electronically, there were always cases when several versions were in circulation and there was often little clarity as to who had approved a document and which was the authorized version. It has to be noted, though, that the problems described in this section were not specific to the foreign service, as other ministries of the Norwegian government had experienced identical problems in relation to implementation of electronic management of documents and information.[25]

---

[22] See *Bedre Elektronisk Saksbehandling i Utenrikstjenesten—Prosjektplan for BEST-prosjektet* [Better Electronic Management of Documents in the Foreign Service—Project Plan for the BEST-Project] Oslo: Ministry of Foreign Affairs, April 30, 2003 (internal document), pp. 3–4.

[23] Based on 2003 interviews.

[24] Based on 2003 interviews.

[25] See *Elektronisk Saksbehandling—Statens generelle kravspesifikasjon* [Electronic Management of Documents—General Governmental Specification of Requirements] Oslo: Statskonsult, 1997.

The situation at the Norwegian missions abroad was no different. As the Norwegian embassy in one of the major North American capitals reported back to headquarters in 2001:

> Although we have installed new IT-systems, we keep the old ways of performing our tasks. This means that instead of capitalizing upon the new opportunities we have only increased the workload, made ourselves more vulnerable, increased the complexity in the management of documents and decreased the quality of outputs. The situation is so precarious that it needs to be taken seriously. We have got fine [IT] tools but cannot use them. We need to get suggestions, instructions, procedures, routines to use the new tools. Nothing will solve itself automatically. We are waiting for something that will make us aware of the possibilities [to use the new IT tools].

To address the problems throughout the foreign service with what one director had called "an unmanageable overload of e-mail and other electronic information",[26] the Ministry has decided to develop its own specific system for electronic management of documents which, as one of the senior officers has put it,

> would meet the specific needs of the foreign ministry and the embassies, which differ from the work of other governmental agencies.

A new system called *BEST*[27] was developed in cooperation with a private software company in 2003. Similar to the DFAIT *InfoBank* system, the idea was to have all software programs used by the foreign service officers in their daily work integrated into a unified user interface. The system was to be connected to a shared electronic archive of the foreign service, so that documents produced or e-mails sent would be easily filed into a shared database accessible throughout the foreign service. Despite the fact that the rationale of introducing BEST was to have a system that would reflect the special needs of diplomatic work and diplomatic communication, the system was pilot-tested at headquarters in the Multilateral Department and in the Secretariat of the Minister of International Development, which are both organizational units dealing with non-traditional diplomatic agenda such as poverty reduction, international environmental issues, distribution of humanitarian aid etc. These units were chosen to test the system, because their staff,

---

[26] Based on 2003 interviews.

[27] This is an acronym for *Bedre Elektronisk Saksbehandling I Utenrikstjenesten* (Better Electronic Management of Documents in the Foreign Service).

contrary to the staff in the more traditional bilateral departments, have been used to a fairly high level of information sharing within the respective unit and across unit boundaries. However, already in the early phases of the pilot project, these units perceived the system as being cumbersome and as slowing down the pace of work. As one director had observed:

> One is used to exchange information and documents simply by using Outlook [a commercially available e-mail program] and this new system is much slower and one needs to learn new routines to operate the system, which in fact makes the system a hinder to effective work in the first phase.

Besides these technical problems, the system featured a set of functions, which basically *re-introduced the principles of bureaucratic hierarchy in the internal electronic communication* by using software called DocuLive. In the system, directors would receive documents relative to particular tasks and they would then electronically distribute them to individual desk officers in their unit, who would hence be put in charge of administering a particular file. Desk officers would then produce appropriate outputs and submit them back electronically to their closest supervisor, who would then send them further up in the hierarchy for capping off or, if appropriate, approve their sending to the respective recipients outside the Ministry. The February 2003 official guidelines for electronic management of documents in the foreign service describe the procedures in detail:

> A director receives a draft of an issue-file from a desk-officer in the electronic in-tray [dedicated for that purpose…]. The director needs to control the document-card details [metadata about the file] and has then three possibilities:
>
> - Administer the file himself/herself, possibly include endorsements or changes. In that case one clicks the 'Administer' button. If this is a paper-letter or fax to external recipients which the desk-officer or secretary therefore have to print out and get the necessary signatures, one uses the 'Forward/Return' button. The document is hence returned to the respective officer who will print it out with a comment that this should be done;
> - Approve the document, but send it further to the next leadership level for capping off, eventually with endorsements and/or changes. In the latter case one clicks the 'Forward/Return' button.;

- Send the document back to the desk-officer with electronic endorsements or changes. In that case one clicks the 'Forward/Return button'.[28]

In the system, each e-mail message or electronic document featured a clear indication of who approved it or capped it off. Hence, as one of the directors observed, "hierarchy had become much more visible".[29] While it was viewed positively by most of the recipients that hierarchical lines of authority were more clear in the system, thereby decreasing ambiguities about accountability, the procedures were viewed as cumbersome. As one director observed:

> this is another step back, because using Outlook, routines were much more informal and flatter, but the new system is again too hierarchical and non-flexible.

The tight following of ideal type bureaucratic procedures which the BEST-system had been imbued with was apparently inconsistent with the usual working processes and communication patterns applied in the daily administrative workings of the foreign ministry. As another informant pointed out, the BEST-system was

> not reflecting the actual ways of how we work in the foreign ministry, which involves lots of horizontal communication and coordination at preparatory stages of decisions.[30]

A further problem with the BEST-system was that the users were each given particular roles in the system, which created ambiguities about how these roles related to the actual roles of officers. There was a distinction between the *role of the leader* and the *role of a desk-officer*. A "leader" in the electronic system had a right to distribute documents and assign tasks. A desk-officer had a different set of rights, mostly to administer files that were assigned to him/her by a "leader". Yet the problem was, that a "leader" role in the electronic system could be given not only to actual leaders of units but also to secretaries, who then had the possibilities to distribute tasks. On the other hand, desk officers holding the role of a "desk officer" in the electronic system could

[28] *Rutiner for elektronisk saksbehandling i utenrikstjenesten* [Routines for Electronic Management of Documents in the Foreign Service] Oslo: Ministry of Foreign Affairs, 10 February 2003, (internal document), p. 7.

[29] Based on 2003 interviews.

[30] Based on 2004 interviews.

not electronically distribute tasks and access information on par with "leaders". This caused problems in situations where actual desk officers were put in charge of coordinating a particular project or a particular cross-unit agenda. The electronic version of bureaucracy and roles of officers in it, which the BEST-system represented, hence hampered the effectiveness of actual working processes, which evolved around more flexible shifting of roles in relation to various situations. Officers at the Ministry were hence forced to go "outside" the system in order to get things done quickly and efficiently, get access to relevant information needed for the execution of tasks etc.[31]

According to the BEST-project plan, the system should have been rolled out at the headquarters by the end of 2003 and in all Norwegian missions abroad by the end of 2004.[32] Yet, given the problems that were experienced in the pilot-phase and throughout evaluations in the early months of 2004, implementation of the system was stalled and eventually abandoned in the Summer of 2004. The substantial investments that were made into the development, pilot-testing and training were lost, but, as one of the respondents argued,

> it takes a lot of guts to take such a step. But if we would have gone on implementing the BEST system despite the problems that it featured, it would have cost us much more in the long run, because we would be stuck with an inefficient system.

In the period November 2004–March 2005, a new comprehensive software solution for electronic management of documents called *eUD Public 360*[33] developed in cooperation with the very same company that developed the BEST-system, was pilot-tested in some of the same organizational units that had pilot-tested the BEST-system. There were at least six factors that could eventually contribute to the success of the new project. The first five concern the differences in the implementation strategy, and the last factor is an environmental change.

[31] Based on 2004 interviews.

[32] *Bedre Elektronisk Saksbehandling i Utenrikstjenesten—Prosjektplan for BEST-prosjektet* [Better Electronic Management of Documents in the Foreign Service—Project Plan for the BEST-Project] Oslo: Ministry of Foreign Affairs, April 30, 2003 (internal document), p. 17.

[33] This means "elektronisk UD" [electronic MFA], while "Public 360" is a specific software platform for electronic management of documents adopted by the Norwegian government in 2004.

The first factor was the change the name of the project. Such a change was necessary, as the project leader in charge of the new project had pointed out,

> in order to get rid of the somewhat negatively charged connotations that the old project name evoked. It was also important not to use a name such as "BEST", which had only raised expectations to unreasonably high levels. This time around we merely point out that this is about electronic foreign ministry and promoted the name of the software product that would be implemented.

The second factor at play was that while the BEST-project was led by a project manager hired in specifically for that purpose from another governmental agency, the new project was led by an experienced foreign service officer.

Having a foreign service officer with extensive contacts throughout the organization in charge, enabled the third factor—a more disaggregated approach to testing and promoting the new software package. Given his personal contacts throughout the foreign ministry, he managed to get a group of 10 experienced foreign service officers located in various departments to build a task-force, who were to devote 1–2 days a week in the pilot-testing period (November 2004 to March 2005) for working with the project, creating awareness of the project in their respective departments and sections, searching for ways of how to adapt the features of the system to the needs and working manners of foreign service officers. Such preparatory work would, as the project manager argued,

> increase the chances for the project to be positively perceived and anticipated among employees as something that will make their daily work more effective.

A fifth factor that seems to have made a difference in the implementation strategy was that the project team had been consciously directing its efforts at middle-level managers throughout the foreign ministry in order to ensure their positive reception of the new software package. As the project manager argued,

> [the middle-managers represent] a group that in some years will be a level higher up in the Ministry [at senior leadership positions] and if we can secure their support for this project today, one can anticipate that IT-supported ways of going about our work will be better anchored in the senior leadership of the Ministry in the future.

Finally, the environmental change that might have contributed to the increased chances of successful implementation of the *eUD Public 360* project was the decision by Statens Forvaltningstjeneste [Governmental Administrative Services] to introduce a shared standard for regulating access to official electronic documents in governmental agencies. As the project leader of *eUD* had argued,

> this piece of legislation finally allowed us to implement the relatively liberal approach to information access that we in the foreign ministry had long been wanting to implement.

This government-wide change in legal regulation enabled the foreign ministry to fully capitalize upon the electronic archive solution in place. As the so called need-to-know principle no longer had to be applied, the vision from the EFS-Report could finally be realized that a pool of unclassified documents would be openly available for access by all foreign service officers irrespective of geographic location. While the implications of this development for the application of the secrecy norm throughout the foreign service are discussed in more detail below in part 5.5, here it can be noted that this allowed for the *flattening of hierarchy at the operative level* as officers at any level now have a searchable electronic archive at their fingertips. The informational asymmetries that usually characterize traditional bureaucratic systems are thereby diminished. The question remains, of course, to what extent N-MFA will experience the same problem as DFAIT with extremely low volume of documents filed in the shared electronic archive.

At the same time, *hierarchy was strictly maintained at the level of decisions*. This was due to the fact that documents still needed to be capped off and approved by superiors, which had not changed with the introduction of e-mails and other forms of electronic information exchange. IT here merely made the processes quicker, while the ease with which the documents were moved in the electronic networks in fact increased the pressure upon the higher echelons. If IT hence had led to changes it is mainly in the volume of e-mails that senior officials in the bureaucracy need to deal with, which is similar as in the case of DFAIT.

Somewhat in contradiction to the latter point but still contributing to the explanation of the strict maintenance of hierarchy at the level of decisions is the fact that the *most senior political figures in the Ministry continued to operate paper-based* at the time of the fieldwork. Any memo or document that was to be approved by the senior political leadership of the Ministry had to be submitted to them on paper. Documents were hence forwarded electronically through the hierarchical levels to the respective secretariats of senior political leaders in the Ministry. There they were printed out and submitted to the senior political leadership on paper, who reviewed and commented on them on paper. Their handwritten notes on the paper documents were then re-typed into the 'grey field for remarks' in the electronic version of the document by the secretariat of the respective senior officer and hence sent back to the organizational unit in charge for revision and further editing.[34]

Finally, it should be noted that the principles and practice of electronic management of documents that were described here concern only unclassified documents. Classified documents still have to be handled mostly on paper with the corresponding paper-based routines.

5.4.2 *Challenges to the hierarchy between headquarters and missions abroad*

Virtually all embassies that participated in a survey on the use of IT conducted by the IT-Section at headquarters in 1999 and 2000 had expressed the need to get access to databases of the headquarters to support their working processes. In its report, the Norwegian embassy in one of the major European capitals held that

> the senior leadership [at the foreign ministry] has to be given the opportunity to understand the [IT-related] possibilities to change the ways we work throughout the foreign ministry as a whole. [One can] create teams, embassies have to be an integral part of the functional departments/desks [at the headquarters], tasks can be delegated to be performed at embassies. This requires access to databases.

Following the switching of networks in 2000, the technical aspects for the informational integration of embassies and headquarters had been laid. *UDintra*—the intranet solution is now available at all embassies. The system features a shared user interface for both the Ministry and missions abroad, including standardized software for electronic management of documents (DocuLive). As pointed out in the internal *Routines for Electronic Management of Documents* of February 10, 2003:

> The routines are worked out with the aim of implementing DocuLive throughout the foreign service, i.e. in the foreign ministry as well as

---

[34] *Rutiner for elektronisk saksbehandling i utenrikstjenesten* [Routines for Electronic Management of Documents in the Foreign Service] Oslo: Ministry of Foreign Affairs, 10 February 2003, (internal document), p. 8.

> at the missions abroad. Therefore, to describe departments, sections, embassies etc., the general terms *organizational units* or *units* will be used. [...] Documents and files sent within the foreign ministry and between the ministry and missions abroad are considered internal memos." (bold letters in the original).[35]

As one of the diplomats stationed on the North American continent observed, the availability of e-mail had been a major step towards integrating informational resources across time zones. While previously, embassy staff would need to send faxes or get up early or stay up late to make necessary phone-calls when particular informational resources were needed, one can now send e-mails, which the individual recipients find in their mailboxes and send back the requested information electronically, so that embassy staff can use them immediately.[36]

When it comes to global collaborative efforts through virtual teams featuring officers at headquarters and embassies, some initial pilot projects have been conducted. One of the first examples was the work on the Norwegian USA-strategy in 2001, where experts from headquarters and staff at the embassy in Washington have been involved using a dedicated intranet-site, where working versions of documents, think-pieces and project-related messages were posted. Yet, as one of the senior officers involved in the work of the virtual team had reported to me, the Intranet-site generated only fairly limited response within the Ministry and the most decisive steps towards coordination and collection of inputs were in fact taken during personal meetings of the various actors involved in developing the strategy. Therefore the officer had to fake several trips to the embassy in Washington and several coordination meetings were held at the headquarters in Oslo.[37]

A further project using the global network connections to integrate embassies with headquarters was the so called *UN-database*, which the Ministry maintained on UDintra during the Norwegian tenure as a member of the UN Security Council. The database featured a pool of unclassified documents and other material relevant to the Norwegian

[35] *Rutiner for elektronisk saksbehandling i utenrikstjenesten* [Routines for Electronic Management of Documents in the Foreign Service] Oslo: Ministry of Foreign Affairs, 10 February 2003, p. 3.

[36] Based on 2003 interviews.

[37] Based on 2003 interviews.

activities at the UN and was accessible throughout the entire foreign service on an open basis.[38]

The fact that embassies now get access to most informational resources available to officers at headquarters enables embassies to perform the strategic analytical work which would have previously been carried out at the headquarters. Hence, for instance the Norwegian embassy in Ottawa performs most of the analyses related to Canada and staff at the corresponding geographic unit at headquarters are used

> to test the ground at home in Norway in relation to relevant actors—public, NGOs, private companies—to find out to what extent certain policies in relation to Canada would be viable on the home-front.[39]

A similar situation was observed by another officer in relation to embassies in Spain, Italy and Portugal, which featured most personnel with expertise related to these countries, while there were only "a couple of officers" with such expertise at the respective geographic desk at headquarters. Connections to UDintra enabled embassy staff to get fully involved in processes which would have previously been performed at headquarters.

Furthermore, the foreign service wide informational integration also enables *unprecedented patterns of informational exchange between Norwegian embassies*. An example of this was for instance when the Norwegian embassy in Beijing sent a memo to the Norwegian embassy in Ottawa on the details about the human rights related meeting between officials from the then Canadian Department of Foreign Affairs and International Trade and the Chinese authorities. That information proved useful and valuable in relation to the dialogue on the coordinated efforts to promote human rights issues that the Norwegian embassy in Ottawa and the Canadian government have been involved in.[40]

Finally, the on-line connections between embassies and headquarters enable horizontal coordination and information sharing among vice-directors at the Ministry, which is supported by a password protected site on UDintra. Vice-directors who are relocated to embassies continue to have access to this tool.

[38] For reasons discussed below in the part 5.5, use of this database tool was eventually abandoned in 2003, to much puzzlement among the Norwegian diplomats.
[39] Based on 2003 interviews.
[40] Based on 2003 interviews.

Through the extension of the electronic networks throughout the foreign service, the traditional division of labor between headquarters and missions abroad is no longer that pronounced. Although it is still only in exceptional cases, rotational personnel with special expertise can continue being involved in specific processes irrespective of geographic location. In particular embassies, where there is a high concentration of country-specific expertise, analytical functions and strategic decision-making migrates from the respective units in headquarters out to the embassy. Possibilities for operative information exchange between embassies have increased. Given the informational integration through electronic networks, the Norwegian foreign service had acquired the technological means to work as a single global organization rather than a collection of organizations connected to the headquarters in Oslo. To what extent this is actually being done is a different matter. As the Minister Counselor at the Norwegian embassy in Ottawa pointed out:

> Although one actually does not get to use most of the documents available on UDintra, the fact that one has the same user interface on the PC and access to most of the same electronic information resources as when one is located in Oslo, makes one feel more integrated and more 'at home' even though one is on a different continent. One is no longer so isolated as before.[41]

Contributing to the sense of being integrated into what is going on 'at home' has also been a small editorial staff at the Department of Press, Culture and Information in the headquarters, whose task it is among other things to post on UDintra short and informal articles about daily occurrences at headquarters.

The level of IT-enabled task-sharing between embassies and headquarters should not be overstated, though. There are limits when it comes to the actual ability of embassy personnel to get involved in cooperation processes with headquarters. As several of the respondents had observed, staff at embassies have simply too many other involvements on the ground where they are and can hence only to a limited extent contribute to projects at headquarters despite the available technological connections. This may vary from person to person, though, and as one of the informants observed,

[41] Based on 2003 interviews.

> the cost of stationing an employee at an embassy abroad is usually three times higher than that of having an employee in Oslo. The foreign service should therefore make sure that these people out there are properly employed.[42]

A final factor worth mentioning in relation to the impacts of IT on diminishing the differences between work at missions abroad and headquarters is somewhat indirect and has to do with the working habits one traditionally had while stationed at home versus the ones abroad. In simplified terms, officers at embassies are used to working "in the field", outside the office, at meetings and conferences, often irrespective of working hours. This is in sharp contrast to the working habits at headquarters which traditionally resemble a standard governmental bureaucracy to a much greater extent—standardized working hours, clear distinction between the office and private premises, work being performed mostly inside the Ministry building etc. This distinction in working habits is being somewhat leveled out by the impacts of IT. An increasing number of officers at headquarters now have *home-office solutions* via desktop PCs and laptops. While this previously had been perceived as a security risk and not encouraged, N-MFA had in recent years been more open towards allowing its employees to use mobile computer connections to work from home or other place. In 2003, more than 100 laptops with the appropriate protection software and communications-tools were distributed among (mostly senior) officers throughout headquarters.[43] By logging on to the system of the Ministry, officers can now report to work irrespective of where they are physically located. As one of the directors observed, this kind of flexible solutions are consistent with the professional culture of the foreign service in general, where rotational officers can hence take back at least some part of the lifestyle and working habits that they usually have while stationed at missions abroad.[44] On the other hand, the on-line connection of embassy computer systems to the databases at headquarters and the increased integration of embassy personnel in collaborative working processes with staff at headquarters, increase the likelihood that embassy personnel have to spend time working in the

[42] Based no 2003 interviews.
[43] Based on 2003 interviews.
[44] Based on 2003 interviews.

embassy offices, using the broadened menu of electronically available informational resources.

In sum, IT implementation had enabled greater information exchange across formal boundaries of organizational units at the headquarters and across geographical distances enabling greater informational integration between embassies and units at headquarters. However, standards for electronic management of documents have largely reflected paper-based procedures, and principles of hierarchical line-authority were re-introduced in electronic information exchanges. These findings match closest with the *renewal scenario*.

5.5 *IT-effects on Secrecy*

5.5.1 *Approach to electronic risks*

Following the 'switching of networks' in 2000, there was a gradual change in the approach to electronic risks at the N-MFA. One has moved away from risk avoidance and towards *risk management*—the realization that not being connected to the Internet constituted greater risk than losing particular pieces of information.

5.5.2 *Policies of internal access to unclassified information*

The switching of networks was consistent with the basic strategic approach to information management introduced in the 1999 report The Electronic Foreign Service (pp. 27–28). This approach stressed the need to increase the protection of classified information and liberalize access to unclassified information. The basic rationale was that with an increasing need for quick reactions to diplomatic events, and with the increased volume of non-traditional diplomatic agenda cutting across traditional organizational jurisdictional areas at the foreign ministry, there is a need for a searchable central electronic archive providing quick and open access to unclassified information and documents. The idea was to have a pool of unclassified documents available in a central electronic database openly available to all employees of the Norwegian foreign service around the world.

The first pilot project using this approach was the above mentioned *UN-database* providing foreign-service-wide access to documents concerning Norwegian activities in the UN. A number of senior and junior diplomats I interviewed thought that this solution was practical and effective facilitating fast access to information and this solution was seen as a model for the Ministry's information access policy, in particular with respect to the planned roll-out of the BEST-system.

In the BEST Project Plan[45] one had suggested a two-tiered electronic archive solution for unclassified documents. There would be the so called *saksarkiv* (issue-archive), where official issue-files with archive-value as specified in the Freedom of Information Act ('Offentlighetsloven')[46] would be kept. Secondly, there would also be the so called *dokumentarkiv* (document-archive), where all incoming and outgoing unclassified documents would be filed, as well as all other documents and memos produced by the foreign service. It was also suggested that working versions of official issue-files and documents should be stored in the document-archive. This would be an openly accessible pool of electronic unclassified information that all employees of the foreign service would be able to draw upon in their daily work. The idea was that every employee of the foreign ministry would be able to decide, when a document that s/he had produced would be filed in the *dokumentarkiv* or *saksarkiv* and hence released for open access to the rest of the foreign service.[47]

However, senior administrative ranks at the Ministry had expressed concerns and were not willing to support this open access solution. The problem was that neither the UN-database nor the proposed two-tiered solution for information access in BEST did justice to the so called *need-to-know principle* that had traditionally characterized access to both classified and unclassified information at the foreign ministry. According to this principle, a foreign service officer only has access to

---

[45] *Bedre Elektronisk Saksbehandling i Utenrikstjenesten—Prosjektplan for BEST-prosjektet* [Better Electronic Management of Documents in the Foreign Service—Project Plan for the BEST-Project] Oslo: Ministry of Foreign Affairs, April 30, 2003 (internal document), p. 12

[46] *'Offentlighetsloven'* (Freedom of Information Act) has been valid in Norway since July 1, 1971. Based on the analysis of Høgetveit (1981), this law applies only to Norwegian public agencies of the central government (ministries, directorates, and various committees), of the local governments, as well as to Norwegian embassies and delegations located outside Norway. This implies that the law does not cover private enterprises, organizations and institutions. However, there are further limitations. The law is only applicable, if a public agency acts on behalf of the Norwegian state or one of its regions (Høgetveit 1981:85). This means that the law is not applicable if a public agency acts on behalf of private, foreign or international organizations.

[47] Files related to personnel-management and consular issues would constitute an exception and not be openly accessible.

information and documents within his/her specific jurisdictional area and only in special cases beyond it (Høgetveit 1981:37–39). The reasons for wanting to keep up this principle in relation to unclassified information and in the electronic environment were, as one senior diplomat explained in 2003 that:

> even unclassified information can represent a security risk if pieces of information from several unclassified documents are combined in the right way.

As virtually all of the informants reported to me, the plans of the administrative leadership to re-introduce the need-to-know principles to regulate access to unclassified electronic archives of the Ministry were negatively perceived among the body of foreign service officers. The UN-database and similar solutions enabling open foreign-service-wide access to unclassified information had been largely perceived as providing the Norwegian diplomats with an informational edge enabling them to effectively live up to their role as central coordinators of Norway's diplomatic activities. Hence, while IT on the one hand supports the Norwegian diplomats in living up to their institutionalized role as central coordinators of Norway's foreign policies, it is on the other hand also prone to undermine one of the central principles upon which the Norwegian foreign service has been operating—the need-to-know principle ensuring maintenance of secrecy. This was the ground around which an *institutional collision* had emerged. While protection of information and secrecy may be central diplomatic norms, so is also the ability to lead in the conduct of the country's foreign relations. IT enactment has in this case facilitated the latter and a conscious attempt to re-introduce the need-to-know principle into electronic management of documents was being met with frustration and negative reactions among the diplomats both at headquarters and at embassies. This was *a collision between two core organizing principles of diplomacy—secrecy and hierarchy*. In a rapidly changing information intensive environment, maintenance of traditional paper-based standards of secrecy was seen as undermining the foreign ministry's ability to maintain its position as the top coordinator of foreign affairs.

Eventually, the institutional collision had been remedied by external government-wide developments, where in the fall of 2004, the Ministry of Modernization introduced new guidelines for regulating access to electronic information in governmental institutions. This new set of guidelines allowed for a more liberal set of executive routines to be developed at the N-MFA and access to unclassified documents was at the time of writing close to the vision presented in the 1999 Electronic Foreign Service Report—an electronic pool of openly accessible unclassified documents searchable by keywords. Only three kinds of unclassified documents are for obvious reasons not openly accessible—a) personnel files; b) consular files; and c) NATO-files, all of which can only be accessed by authorized personnel. Senior administrative leaders at the Ministry had according to one informant

> understood that there is no way to capitalize upon the new technological solutions in place if we keep on applying the need-to-know principle.

The logic behind the decision was apparently that,

> at the N-MFA the need to handle large amounts of complex information quickly is now much greater than the need to protect information.

The fact that all documents that are saved in the electronic archive are now openly accessible throughout the foreign service had brought about new challenges for individual foreign service officers. As the need-to-know principle had practically disappeared in the management of unclassified documents, individual officers, as one of them had argued, experience the "*need to hide*" documents. As the officer further elaborated, this relates to documents that are unclassified but contain sensitive situational information that might need to be protected from public access, or also to "documents that are work-in-progress and it would do more harm than good if people would use the information".[48] More generally, a situation when officers are individually responsible for protecting documents they do not wish others to see might eventually lead to a decrease in the volume of documents filed in the central electronic archive and data-hoarding on individual PCs—a situation that the Canadian foreign service had experienced in relation to its openly accessible electronic archives (discussed in more detail in the previous chapter in part 4.4.1). This would deepen the existing fragmentation of the institutional memory.

The fact that access to unclassified information was liberalized does not mean, however, that secrecy as a norm would be disappearing. Information at the lower levels of classification is exchanged within the Blue network and highly classified information is not in electronic form.

---

[48] Based on 2004 interviews.

As for the more general rationale behind the need to keep secrecy as a norm in daily diplomatic business, a senior diplomat pointed out that

> Usually, the Norwegian foreign ministry does not feel the need to withhold information from the Norwegian public. However, we need to be reliable partners for other foreign ministries entrusting us with classified information. They need to be able to rely upon us. [...] And besides that, protection of information and of its sources is an important characteristic feature of our work. This is what our work is about.[49]

This statement indicates the institutional pressures of the global diplomatic community distributing professional standards and norms across the collection of foreign ministries around the world. Protection of information and maintenance of secrecy as a norm hence is not necessarily the property of any individual foreign ministry, but of the field as a whole. The last sentence indicates that secrecy and protection of information are important elements of a diplomat's professional identity, which forms the basis of the logic of appropriateness informing his/her actions. This indicates the nature of diplomacy as an institutional order. The core organizing principles of diplomacy—including secrecy—are maintained by their deep institutionalization in each individual foreign ministry. In other words, this is an institutional constellation that ensures national organizational maintenance of a set of transnational organizing principles. I will return to this in the theoretical discussion in the concluding chapter in part 7.4.2.

Coming back to the findings, the latter quote is also directly related to the diplomatic culture at the N-MFA, where having access to information (classified or unclassified) that others in the organization do not have access to, has been, as one director had pointed out, an important *status symbol*. The fact that a particular document is declared classified provides the document with a special aura of importance. Yet as the director further noted, classification of documents is conditioned by the specific situation in which documents are declared classified, which has a time-dimension attached to it. As he explained,

> what is secret today, may be public or unclassified information tomorrow. Classified information is very often 'fresh-ware'.

Overall, internal information access at N-MFA is no longer being policed. In the quest for higher effectiveness of analytical work, one has

[49] Based on 2003 interviews.

opted for an open-access solution to a pool of unclassified documents, while classified documents remain accessible only through specialized PCs or on paper. This conforms to the notion of *guarded openness*, which suggests a match with the *renewal scenario* in the analytical framework developed above.

5.6 *IT-effects on Communication with the Public*

Due to the general skepticism of the Norwegian public towards the N-MFA, the traditional cautiousness of the ministry in handling public information had always been complemented by a "need to share information with the public" in order to get public support and approval of foreign policy activities.[50] Internet had become an effective vehicle for such communication activities. Ever since the Norwegian government established its official web-sites, the web-site of the foreign ministry has been the most extensive one in the context of the Norwegian government in terms of the amount of documents posted and in terms of the number of visitors. The fact that many of the documents were produced in several language versions further widened the volume of material available on the cluster of web-sites operated by the Ministry. In addition to the rich set of documents, the web-site featured videos from conferences and audio-files. Web-sites of all Norwegian embassies got a standardized design and informational architecture, and were connected to the so called Norway Portal[51] introduced in late 2003. The portal was considered "Norway's official face to the world" and received a high-quality web-design, for which it received the Good Design 2004 award of the Norwegian Design Council. Every embassy web-site on the Norway Portal has culturally-specific features (locally-relevant information, sometimes local language etc.). An interesting and innovative feature of the embassy web-sites is a set of around 40 web-cams placed out around Norway, so that visitors can take a peak into Norway through the embassy web-sites, which is an entirely new way of direct real-time country presentation.

---

[50] Based on 2003 interviews.
[51] http://www.norway.info/.

5.6.1 *Direction of communication*

Although the quality of the design of the N-MFA web-site, of the Norway Portal and embassy web-sites is high, the N-MFA had not been encouraging the general public to voice opinions on foreign policy and the pattern that the publishing of documents on the web-sites followed had been that of *one-way communication*. The general rule of thumb practiced throughout the foreign service had been that only documents, reports and memos published on paper were to be published on-line.[52] This does not, however, mean that N-MFA does not have two-way contact with societal actors. The most usual and well-established pattern of public consultation is that societal actors involved in particular agenda are regularly called in to committees and other meetings with representatives of the N-MFA. This relates for instance to consultations and cooperation with NGOs on developmental issues, consultations of various interest groups in relation to for instance WTO negotiations etc. The work of such formalized committees is supplemented by continuous IT-supported information exchange between N-MFA units and the specific societal actors. As one of the senior N-MFA's officers pointed out in 2003,

> practically all officers now have a network of contacts outside the Ministry with whom they regularly communicate by e-mail. This is very different than 15 years ago when there was no e-mail and all written documents that were sent outside the Ministry had to be capped off by senior officers in each respective unit. Today, the walls of the Ministry are definitely more porous.

Following the 2004 integration of NORAD (Norwegian Agency for Development Cooperation) into the structure of the N-MFA and the integration of *Innovation Norway* representatives into embassies, the N-MFA is more likely to have increased and more regularized consultation mechanisms with on the one hand NGOs and academics regarding development issues, and on the other hand the business community concerning export activities. In addition to this, through the amalgamations with NORAD and Innovation Norway, N-MFAs channels for communication with external actors have multiplied as have communication styles that these new actors bring into the Norwegian foreign policy network that was created through the amalgamations.

[52] Based on 1999 and 2003 interviews.

Through the amalgamations, the need for involving citizens in consultation on foreign policy issues is likely to increase and not diminish, because while the N-MFA was on its own it could to a greater extent argue that it administered high politics, which ordinary citizens had little understanding of. As NORAD and Innovation Norway make issues of low politics an integral part of the N-MFAs agenda, there might arise a greater need for increased public consultations. At the time of writing, though, N-MFA's communication with the general public was conducted in a one-way and ex-post fashion.[53]

5.6.2 *Level of centralization of public communication*

N-MFA's communication with the public had been highly centralized with information by and large provided either by the Minister or the spokes-person of the Ministry. The latter had a small staff of information officers working under him at the time of fieldwork in 2003. The Department of Press, Culture and Information (PCI) had been handling most of the Ministry's external public communication. The N-MFA has also been administering *Norway's International Press Centre*, where it provided free-of-charge office space, logistical and IT-support to foreign journalists stationed or on assignment in Norway. Also, a *web-based Press Service Center* has been established, where journalists could register on-line and receive accreditations to particular media events organized by PCI and other updated information on changes in schedules, upcoming events etc.[54]

While N-MFA's communication with the general public remains centralized, IT solutions open up the possibility for individual units within the headquarters working on a particular agenda, to create *virtual project rooms* outside the firewalls protecting the N-MFA communication networks and hence engage in regular information exchange and communication with external actors. The first such virtual project room was used for cooperation with the private company that has been developing software solutions for the N-MFA in recent years

[53] The Asian tsunami-catastrophe in December 2004 exposed N-MFA's total lack of procedures for handling a crisis-situation volume of information-requests from the general public. See *26.12. Rapport fra evalueringsutvalget for flodbølgekatastrofen i Sør-Asia* [26.12. Report by the Evaluation Committee for the Tsunami-Catastrophe in South-Asia] Oslo: Statens Forvaltningstjeneste, April 2005.

[54] See http://www.regjeringen.no/nb/dep/ud/Pressesenter.html?id=836.

(BEST and most recently eUD Public 360). Templates and Blogs of web-solutions, web-portals and other contributions to the project-work have been posted there. Such virtual project rooms were likely to be used for enhanced cooperation with other more directly policy-related actors such as NGOs, academics or other governmental ministries.[55] Communication with the public/actors outside the N-MFA would hence take the shape of pre-defined project issue areas, for which protected virtual rooms could be used. While the IT-supported format of such communication would be innovative, the activities as such would be a continuation of existing committees and other consultative mechanisms with external actors as described in the previous section.

5.7 *Conclusion*

In response to a perceived performance crisis, N-MFA had conducted a radical modernization of its IT-infrastructure following the recommendations of the 1999 Electronic Foreign Service Report. IT had enabled unprecedented information exchange across organizational unit boundaries and across layers of hierarchy, in particular at the operational level. Implementation of technological solutions without corresponding update of procedures for electronic handling of documents had however led to severe inefficiencies where IT-based tools were used in support of paper-based procedures or IT-based tools were used without regard to the need for proper record keeping. Procedures for electronic management of documents introduced by N-MFA had reinvigorated hierarchy as a core organizing principle, but also enabled informational integration supporting virtual teams and closer cooperation between headquarters and embassies. In terms of effects of IT on the hierarchy, the evidence from N-MFA hence matches closest the *renewal scenario*.

In relation to the secrecy norm, the Ministry's general approach to electronic risks has moved from risk avoidance to risk management. IT enactment has initially led to an unstructured liberalization of the Ministry's policy regulating internal information access. Although there were attempts to re-assert the need to know principle by the senior ranks in the Ministry's administration, these attempts have been meeting resistance and skepticism among diplomatic staff, who had experienced

[55] Based on 2004 interviews.

the benefits of more open information access solutions. The solution adopted is one of open access to a pool of unclassified information and severely limited access to classified information, which fit well with the idea of *guarded openness*. The scenario that the empirical evidence matches the closest is therefore *renewal.*

Finally, in terms of communication with the public, the Ministry has well developed web-portals providing a rich collection of information related to Norway's foreign relations. Patterns of regular information exchange and consultation have been developed with specific actors and constituencies inside the country. Electronic communications have intensified these activities. Yet communication with the general public has had the character of one-way press-releases, where the Ministry has been communicating as a coherent whole—usually through political leaders or the spokes-person. In relation to the analytical framework developed in chapter 3, this suggests a match with the *status-quo* scenario.

If we agree with the above mentioned notion of Knudsen (1997), that foreign policy formulation in Norway had shifted from the foreign ministry out towards a societal political space with a multitude of actors, then the inability of the N-MFA to engage in effective consultation with the public leaves the Ministry continuously struggling for gaining societal approval and legitimacy for its activities. A new process of organizational modernization initiated by foreign minister Gahr Støre in March 2006 promised to deliver fewer hierarchical levels in the organizational structure; multiplication of channels of external communication to include the ministry leadership, the press-spokesperson and respective experts from within the bureaucracy; and an improved technological integration of informational resources in order to develop a global knowledge-organization.[56] These reforms follow up on the recommendations of the 2000 Electronic Foreign Service report and further research would be needed to assess their possible effects.

The next chapter addresses the same processes in a foreign ministry founded amidst the information revolution—the Slovak MFA.

[56] Jonas Gahr Støre: Utenrikstjenesten—En kunnskapsorganisasjon. (The Foreign Service—A Knowledge Organization), presentation, Ministry of Foreign Affairs, Oslo, March 22, 2006.

CHAPTER SIX

MINISTRY OF FOREIGN AFFAIRS OF THE SLOVAK REPUBLIC

As far as diplomacy is concerned, we didn't know much about it, but we had lots of enthusiasm and took on any task with great verve. At that time, we didn't care how long we would have to spend at headquarters, and when and where we would eventually be assigned to work abroad. We were happy enough to get a couple of computers installed and we were amazed at T-602—then a miraculous piece of software.[1]

6.1 *Introduction*

Compared to many other foreign ministries, the Ministry of Foreign Affairs of the Slovak Republic (MFA SR) is a young organization, established on January 1, 1993, when Slovakia declared its independence and became a successor state of Czechoslovakia. The first years of the MFA SR's operation were characterized by frequent shifts in the political leadership of the organization[2] and by logistical instability.[3] Political change after the parliamentary elections in 1998 brought about a stabilization of the basic axioms of Slovakia's foreign policy and a consequent effort by the new government to "catch up" with the

[1] A Slovak foreign service officer describing the situation at the Ministry of International Relations of the Slovak Republic in early 1991 (quoted in Mojžita 2004:123, author's translation). T-602 was a rather cumbersome Czechoslovak text-editing program used throughout the Slovak government prior to the general spread of software such as Word Perfect and MS Word in the late 1990s.

[2] During the first five years of the Ministry's operation there was a change on the ministerial post on average more than once a year. Hence, when foreign minister Kukan took office after parliamentary elections in 1998, he was the seventh foreign minister the country had since gaining independence in 1993. It can also be mentioned, that only three times the foreign minister was a career diplomat—Eduard Kukan—serving for some months in 1994, and then for eight years between 1998–2006; and Ján Kubiš since 2006.

[3] The headquarters of MFA SR had moved three times between 1993 and 1998, and there was also a number of unresolved issues regarding former Czechoslovak property on embassy compounds abroad.

neighboring Central European countries in the process of integration into the EU and NATO.[4] For the MFA SR, which was put in charge of coordinating the "catching up" process, this involved a somewhat one-dimensional focus and drained away most of its organizational resources. What is more, the Ministry also operated in the context of groundbreaking public administration reforms that were about to determine "what kind of a state the Slovak Republic should be".[5] These reforms along with the EU accession process brought about frequent shifts of administrative procedures, routines and standards used by the Slovak governmental organizations in their daily work.

In light of these circumstances, the strategic use of IT in support of the Ministry's operations was awarded meager attention by the senior political leadership during the first ten years of the Ministry's operation. And yet, IT has been used widely at headquarters and at Slovak missions abroad and numerous innovations were introduced. Effects of IT at MFA SR went hand in hand with the standardization and gradual institutionalization of administrative processes, structures and routines throughout the organization on a global scale, and more generally also hand in hand with the development and introduction of government-wide e-governance practices throughout the governmental administration in Bratislava.

The chapter first outlines the institutional characteristics of MFASR. This is followed by a description of its IT infrastructure. As in the previous chapters, effects of IT implementation and use are then discussed in relation to the core organizing principles of diplomacy. Conclusions follow.

6.2 *Institutional Background and Characteristics of MFASR*

Already prior to the dissolution of Czechoslovakia in 1993, efforts were under way to establish organizational strcutures that would enable the

---

[4] In 1998, the Czech Republic, Poland and Hungary were about to became members of NATO (in March 1999). Together with Slovenia and Estonia, these countries were also ahead of the rest of the Eastern European countries in the negotiation processes leading to the entry of the EU.

[5] *Strategy of Public Administration Reform in the Slovak Republic*, Plenipotentiary for the Reform of the Public Administration, Government of Slovakia. Bratislava: July 1999, p. 3 (available at http://www.mesa10.sk/vs/index.asp?id=1).

Slovak government[6] to assume a greater international role and increase Slovakia's visibility internationally. This has prompted the founding of the Slovak *Ministry of International Relations* in 1990. It was led by Milan Kňažko, a former actor and one of the leaders of the 1989 anti-communist movement in Slovakia, and manned by a heterogeneous group of dedicated officers with varying professional backgrounds. The organization adopted standard bureaucratic procedures and structures usual throughout other organs of the Slovak government, but the diplomatic routines and practices remained mediocre.[7] The character of the organization began to change more substantially in the direction of a standard foreign ministry in the second half of 1992 with the gradual arrival to Bratislava of Slovak diplomats formerly employed at the Czechoslovak federal foreign ministry in Prague who opted for Slovak citizenship in the wake of the anticipated declaration of independence. Their influx involved introduction of standard diplomatic routines, operating procedures, protocol and working processes by and large inspired by their former administrative and diplomatic practice at the Czechoslovak foreign ministry. This led to an increasing infusion of the organization by a more standardized diplomatic culture, albeit applied under new administrative and political conditions. Miroslav Mojžita (2004:85), a senior diplomat who directly participated in these processes described the developments in the following manner:

> ...Of course, our ambition was not to create a one-third-size Černín Palace[8] in Bratislava. We knew that the whole system of procedures, guidelines and organizational structures was tailor-made for another state and indeed for another political system. Moreover, foundations for a modern and effective institution in the form of the Ministry of International Relations were already laid in Bratislava. The 'software' from Prague was hence important merely for comparative purposes and for double-checking of particular processes. And yet, the diplomatic administration has its constant principles and procedures stabilized through centuries [of

[6] Since 1969, Czechoslovakia was formally a federation, where Slovakia had its own government ministries of education, agriculture, industry etc. These had their counterparts in the Czech Republic and at the federal level, where the latter in addition to these responsibilities also managed foreign affairs and defense.

[7] An in-depth look at the emergence of Slovak diplomacy and at processes that guided the formation of the Ministry of Foreign Affairs between 1990–92 is provided by Mojžita (2004).

[8] The *Černín Palace* was the residence of the Czechoslovak Ministry of Foreign Affairs in Prague until 1992. Today it houses the Ministry of Foreign Affairs of the Czech Republic.

> diplomatic practice]. These principles and procedures were embedded very rigorously in the bureaucracy of the Černín Palace, many of them dating back to the Habsburg monarchy. All of this spiritual heritage belonged also to Slovakia and it was of utmost importance to not only move it to Bratislava, but also to fully adopt it and apply it in practice" (author's translation).

Hence, given this preparatory work, when Slovakia became independent on January 1, 1993, the 'new' *Ministry of Foreign Affairs of the Slovak Republic* had the organizational capabilities enabling it to assume the scale of responsibilities that are usually required of a foreign ministry.

6.2.1 *Organizational structure*

Since its founding, MFA SR has been characterized by frequent changes of its organizational structure. While these structural adjustments were often the result of internal turf battles over limited resources, which is a usual occurrence in most public organizations, they were also happening in the context of the above mentioned groundbreaking governmental reforms setting the basic fundamentals of the Slovak public administration.[9] In addition to these administrative processes, fundamental axioms of Slovakia's foreign policy were also being defined which also reflected in changes in foreign policy priorities and recent developments in the foreign policy environment.[10] The most recent changes in the organizational structure of MFA SR were introduced in March 2005 and generally reflected Slovakia's position as a member of the EU and NATO and, more specifically, also reflected MFA SR's position as the central coordinator of the Slovak governmental involvement in the political processes of the EU.[11] Although there were several changes of the organizational structure between the two rounds of fieldwork that

[9] *Strategy of Public Administration Reform in the Slovak Republic* Bratislava: M.E.S.A. 10, published July 1999, p. 3 (available online at http://www.mesa10.sk/en/vs/index.asp?id=1)

[10] A good example of this *ad hoc* strategy was the founding of a special department for legal proceedings before the International Court of Justice in The Hague during the litigation over the Gabčíkovo-Nagymaros Waterworks in the mid-1990s or establishment of the post of a special state secretary for European integration immediately after the 1998 parliamentary elections when it was necessary to boost the new administration's drive aimed at closing the gap behind neighboring countries in the process of EU accession negotiations.

[11] For the organizational structure introduced in 2005 see http://www.foreign.gov.sk/En/organization_chart_no_names_en.pdf

provided data for the current analysis (1999 and 2003 respectively), the main traits remained the same. In 2003, MFA SR was headed by a minister of foreign affairs assisted by two state secretaries and the head of administration, and was divided into six sections (see Appendix 6).

The ministry operated with three basic kinds of organizational units: sections (*sekcie*), departments (*odbory*) and units (*oddelenia*). The latter ones often featured only 1–4 employees and usually served as a means of providing senior officers rotating back to headquarters after assignments abroad with a management position and thus higher salary. This, though, was perceived as an organizationally inefficient solution as department directors could manage 10–12 employees and often tasked line officers directly circumventing the unit-directors.

Abroad, the Ministry operated in 2003 through a network of 76 diplomatic missions (i.e. embassies, permanent missions, regular missions and consulates general),[12] 1 branch office (in Bonn), 8 Slovak Institutes (in Berlin, Budapest, Moscow, Paris, Prague, Vienna and Warsaw),[13] a so called Community House and a Commercial and Technical Center in Moscow, one seasonal consular office operated six months a year in Kuwait during the spa and tourism season,[14] and 102 honorary consulates.[15] In addition to MFA SR officials, trade officers under the jurisdiction of the Ministry of Economy of the Slovak Republic were posted at 47 Slovak missions abroad, where they manned the so called Trade and Economic Departments (*Obchodno-ekonomické oddelenia*). Slovakia hence operated abroad through a bi-cephalous foreign service with separate political leaderships and administrative procedures. Despite their physical co-location at Slovak missions abroad, the two strands of the foreign service often had problems ensuring effective coordination of activities.[16]

---

[12] 32 embassies in Europe, 6 embassies in the Americas, and 20 embassies in Asia, Africa and Australia.

[13] These are institutes in charge of cultural promotion.

[14] This office serves to facilitate the flow of clients from Kuwait and other Arab countries receiving treatment in Slovakia's spa towns.

[15] The number of honorary consulates had increased to 144 in March 2007 and further 11 honorary consulates were expected to be opened throughout 2007. See *Správa o stave siete ZÚSR v zahraničí v r. 2007 a východiská pre jej ďalší rozvoj* [Report on the status of the network of Slovak missions abroad and on possibilities of its further development in 2007] Bratislava: MFASR, March 31, 2007.

[16] This problem was partly related to the fact that as late as in the summer 2007, there still was no formal agreement setting the standards and procedures for

Due to financial restrictions in consecutive state-budgets in recent years, MFASR has been increasingly operating through what the head of the ministry's administration, Peško (2007), had called 'mini-embassies' featuring the ambassador and one other diplomat. A problem related to the work of missions abroad, and a particularly serious one at the 'mini-embassies', has been the excessive amount of time required by the cumbersome administrative routines required in the daily operations of a post abroad. As Peško (ibid.) reported, mini-embassies allocate as much as 50% of their work-time to filling out forms and performing other administrative tasks required by the headquarters, which precludes them from performing more important tasks related to the country of operation. Moreover, due to limited resources and staff, some diplomats are required to cover five to six multiple agendas. An officer at a mission abroad could work on issues ranging from political and economic reporting, through financial management and administration of the post, to consular matters (ibid.). To address these and other organizational challenges both at missions abroad and at headquarters, the MFASR has been involved in a broad-based organizational reform process since the early months of 2007. This involved audits of organizational structures, procedures, work processes, training of personnel and not least the IT-infrastructure.

6.2.2 *Personnel and organizational culture*

Following its establishment in 1993, MFA SR had to hire an extensive number of new employees to ensure the functioning of the network of Slovak diplomatic missions and to assume the full range of responsibilities for the operation of the headquarters. This situation could be considered an organizational shock and during the first year of its existence, there was approximately a four-fold increase in the numbers of MFA SR staff counting about 700 employees by the end of the year. This radical increase was not accompanied by targeted socialization procedures and professional training for newcomers. As a result, the Ministry was at the time of fieldwork still lacking an integral common organizational culture or an *esprit de corps* which characterizes more traditional foreign ministries. Instead, it resembled a conglomerate of

cooperation and coordination between trade officials and MFA SR officials at Slovak missions abroad.

organizational cultures that individual employees brought from their previous professional careers. The relatively small group of career diplomats who formerly served in the Czechoslovak foreign service stood out as the most coherent grouping and has been influential internally occupying about 60–70% of senior leadership posts in 2003.[17]

Besides the 'on the job' socialization that new employees would go through at Headquarters and later at assignments abroad, there was at the time of fieldwork no standardized socialization procedure for new diplomats comparable in character to the *Aspirantemnda* of the Norwegian MFA. This lacuna had *not* been substituted by the Slovak system of higher education, because, at the time of writing, there was no educational institution in Slovakia that would be considered a diplomatic academy. However, the MFA did have an internship scheme for students of some selected universities, who then provided for a large part of newly recruited personnel. In this area, the Faculty of International Relations of the Matej Bell University in Banská Bystrica and the Institute of International Relations and Law Approximation at the Faculty of Law of the Comenius University in Bratislava were the most notable cooperation partners of the MFA in 2003. In addition, the Personnel Department of the MFA operateed a database featuring information on approximately 4000 prospective candidates to foreign service at any point in time. According to specific needs, the Ministry contacts the applicants from the database, as well as other suitable candidates, and invites them for a set of tests and interviews.

The Ministry was providing its employees with some continuous education and training. Yet those were not extensive and for instance only 0.25% of the overall annual MFA SR budget was used for training of personnel in 2001.[18] Some of the expenses related to diplomatic training of MFASR personnel were also carried by the EU and NATO, or by their member states. This included study-stays of Slovak foreign

---

[17] Based on 2003 interviews.

[18] Based on MFA SR information provided as part of the *Survey on Knowledge Management Practices for Ministries/Departments/Agencies of Central Government in OECD Member Countries*, PUMA/OECD, 2002, pt. 1.2.e. Given the fact that the menu of language courses provided by the Ministry is limited, officers have been taking language courses at external institutions, which they often also had to cover from their private funds. First in 2004, the Ministry had decided to partially cover employees' expenses with such external language courses (see *Zameranie zahraničnej politiky SR na rok 2004* [Strategic direction of the foreign policy of Slovak Republic for the year 2004], p. 45 (http://www.foreign.gov.sk/files/zameraniezp_2004.rtf).

service officers at diplomatic academies and other educational institutions in European as well as non-European countries.[19] A further way of acquiring new knowledge and training is for MFASR staff to take on assignments with external actors such as international organizations or NGOs. This kind of external mobility was 3.9% of staff in 2001.[20]

As late as 2007, the MFASR still did not have a defined organizational mission identifying key relationships to domestic and foreign stakeholders, and a core set of organizational values was also missing. Nor was there any clearly identifiable foreign service code of conduct. This could be partly related to the fact that a legal act regulating the Slovak foreign service was still being developed in early 2008.

Due to a relative lack of formal procedures for the horizontal exchange of information between departments (even within the same section), and only limited access to the senior leadership in the ministry, a culture of informal information exchange has been developing within the organization. While this kind of an alternative culture is a common feature in most organizational settings and alternative authority structures can emerge, the primary challenge with this situation in the MFASR is the possibility of loss of time and lack of information when decisions on particular foreign policy issues are to be made effectively. If informal links are missing, documents and information provided to colleagues requesting them are often too 'thick'—containing too much information on too many aspects of a given issue—('just to be on the safe side', as one officer put it) and hence difficult to work with.

By the end of 2002, the Ministry had a total of 1,173 employees, 478 of whom worked at the headquarters in Bratislava, while 695 were posted at the country's diplomatic posts abroad. The Ministry had a relatively young staff as the average age of employees at Headquarters was 39.[21] These numbers were approximately the same in 2007.[22]

---

[19] Among the partner institutions are diplomatic academies in Clingendael, Madrid, Moscow and Vienna. University of Oxford is also among the educational institutions providing study stays for MFA SR personnel. Approximately 5–10% of the Ministry's staff annually participate in these educational visits that are usually financed by host institutions and last from one to twelve months.

[20] Based on MFA SR information provided as part of the *Survey on Knowledge Management Practices for Ministries/Departments/Agencies of Central Government in OECD Member Countries*, PUMA/OECD, 2002, pt. 1.2.k.

[21] Numbers from *Aktuálny ročný komponent výročnej správy MZV SR* [Current Yearly Component of the Annual Report of MFASR], Bratislava: MFA SR, 2003, pts. A 1.4.1 and A 1.4.2.

[22] Peško (2007) reports that the MFASR had 1150 employees in 2007. Of these, 390 diplomats and 280 administrative-technical staff were posted at Slovak missions abroad.

6.2.3 *Domestic environment*

As a relatively new foreign ministry of a newly independent state, the MFA SR was at the time of fieldwork in 2003 still in the process of defining the elementary axioms of its role in relation to the domestic society. This process was facilitated by the fact, that MFA SR has since its foundation been enjoying a high standing and respect throughout broad strata of Slovak society. The conduct of foreign policy had mostly been endorsed by high approval rates in public opinion polls, in particular since 1998, when the country reinvigorated its efforts to join the EU and NATO and has achieved notable progress in a fairly short time (Gyárfášová and Velšic 2002). Paradoxically, the high levels of approval might be related to what Lukáč (2004:319) had called the *introspective nature* of Slovak political elites and society, which have traditionally lacked a broader internationalist perspective that would enable greater societal engagement in matters of foreign policy. Generalizing somewhat, one could argue with Lukáč (ibid.) that the broad strata of the Slovak public had been considering foreign policy to be of such a nature that 'ordinary people' know little about and which therefore should be left in the hands of experts of the MFA SR. The foreign ministry has therefore had the advantage of operating in a domestic public environment debating foreign policy issues to a limited extent and societal actors exercising only limited pressure on the foreign ministry to legitimize foreign policy decisions (Bátora 2003, ibid.). This has enabled the Ministry to start adopting an elitist role as the top coordinator steering the foreign affairs of the country.[23]

The fact that the Ministry has been aware of the gradually emerging need for greater legitimization of its role and performance has been reflected in the internal review of its organizational capacities conducted by an internal task-force in 2002 and 2003 (discussed in more detain in part 6.4.1). These tendencies were strengthened in particular following Slovakia's entry into the EU and NATO in 2004 involving new expectations by partner governments of active Slovak inputs and involvement in international crises and global policies. As Nič (2007:34) argued, such a shift in expectations and aspirations represented a

[23] The MFA SR is aware of this fact and its self-image had reflected it accordingly. As the head of administration of the ministry Mojžita (2004:27–28) held:

> "...a general rule applies: while the bureaucrats at the finance ministry or the ministry of economy might very well have higher salaries, it is always a greater honor to be a bureaucrat at the foreign ministry. [...]"

"mental change" in Slovak politics and a search for various ways of how Slovakia could contribute to external relations of the EU and to NATO's global engagement has been underway. Of key importance in this process has been the work of Slovak NGOs supporting and promoting Slovak engagement in areas such as the Western Balkans,[24] the Ukraine and Belarus, but also in more remote places like Cuba. This kind of "civic diplomacy" (*občianska diplomacia*) proved instrumental in enabling Slovakia to start carving out a niche in overall external relations of the EU. Realizing this potential, the MFASR established an organized platform for coordination and cooperation with internationally active Slovak NGOs in 2004.

6.3 *IT-infrastructure of MFASR*

The build up of the technological capacities of the MFA has been carried out in the context of a government-wide initiative launched in 1993. The thrust of the initiative was to create a shared IT-backbone system for the Slovak government, and a governmental network called GOVNET was launched.[25] It links the Prime Minister's Office, Chancery of the President, the Parliament, all the ministries and other governmental agencies located in Bratislava, as well as state agencies located in the regions of the country. It is used for electronic mail communication, World Wide Web access, provides a web-portal for sending sms-messages to governmental cell-phones, and enables dial-up connection for remote users.

The link-up of the MFA SR to GOVNET was in the first five years of its operation limited due to logistical instability and the frequent relocation of the headquarters, which inhibited build-up of the necessary technological infrastructure at the foreign ministry. Instability was magnified by the fact that some organizational units were scattered in various buildings away from the headquarters, including among others also the then Department for Special Communication (DSC) in

[24] For a useful review of recent Visegrad 4 initiatives in relation to Western Balkans see Strážay (2007).

[25] GOVNET was established according to the governmental resolution nr. 310/1993.

charge of all external communication of MFA SR. In addition to the unstable logistical conditions, diplomatic staff at headquarters had to struggle with purely technical challenges stemming from both available hardware and software: 1) hardware was mostly old and hence entirely inadequate to handle larger amounts of data or access to web-sites; 2) software products in use varied in versions and providers, which often limited compatibility of exchanged files. Given this situation at the headquarters, technological innovations during the first five years of the Ministry's operation came almost exclusively from individual Slovak embassies, which were able to purchase new hardware and software, set up internet connections, as well as in some cases develop individually designed embassy web-sites.[26] These individual micro-innovations at posts abroad contributed to the growing multitude of standards of hardware and software throughout the foreign service, so to facilitate the communication between embassies and the headquarters, the DSC had to convert all files into the electronic formats used by the respective recipients.

Adding to the problematic nature of varying hardware and software in the turbulent period prior to 1998, was the conscious strategy of the MFA tech-support staff who preferred hardware upgrades and purchasing parts to "build" computers on their own, rather then purchase commercially available off-the-shelf workstations.

It was hence not until 1998, when the Ministry moved into its current premises, that a serious build-up of technological infrastructure and a more coherent approach to the management of IT assets could be initiated. DSC had merged with the Department of Information Systems and the *Department of Informatics and Special Communications* (DISC) was created in late 1998. This reorganization was also accompanied by a shifting out of the senior management of the Department and most of its personnel. Hence, more than 80% of the tech-support personnel working at headquarters and the embassies in 2003 had been employed after 1998. The new management of DISC introduced new

[26] The Slovak embassy in Ottawa was among the pioneers in web-presence. As the .com suffix in its former URL (www.slovakembassy.com) well indicates, this was a very early web-site and established on a commercial basis using the services of a local web-provider. As of Fall 2003, all embassy web-sites have a common look, and are operated from a central server at the MFA in Bratislava: www.mfa.sk.

client-oriented service standards, as well as new strategic approaches to IT.[27]

This was firstly related to acquisitions of software and hardware, where the previous procedure of upgrading machines and 'building' computers was abandoned for off-the-shelf commercial solutions. This led to a gradual standardization, so that in 2003, all workstations at headquarters and at Slovak missions abroad used unified types of software solutions. One of the biggest challenges that MFA SR had to deal with in terms of the technological infrastructure build-up was the creation of global online connections between embassies and the headquarters. As of 2003, the MFASR uses an approach similar to the N-MFA and operates *two networks*—one connected to the Internet and used for standard production of documents and daily work by all officers, and a second network unconnected to the Internet, used for classified communication, accounting, personnel systems etc. Virtually all Slovak missions abroad have been connected to these two networks since 2004. In this way embassies perform their accounting tasks connected to the central accounting systems of the Ministry. The same applies to the so called Schengen-database, where consular and visa-officers at embassies can search for visa-relevant information in a central registry. In addition to the enhanced access to certain databases of the Ministry, the on-line connections of embassy-systems with the HQ also enable tech-support and maintenance of electronic systems at embassies to be performed directly by IT-specialists from headquarters in Bratislava. This has brought about an effectivization of the tech-support function both in terms of cutting down on the time it takes to fix a technical problem at en embassy irrespective of its location and in terms of lowering costs as IT-specialists in most cases no longer have to travel

[27] Given the fact that DISC had a permanent shortage of IT-specialist personnel, it was becoming increasingly normal to outsource particular IT-functions to private sector contractors. Such contractors were for instance hired to be at the Ministry one or two days a week at certain periods, when the volume of work to be done was particularly extensive, to support operations of electronic systems supporting accounting and personnel databases. In all, such outsourcing activities accounted for approximately one person-year of salary. Still, it is seen as a cheaper solution than to actually hire an IT-specialist, because there would be additional costs related to training the employee, while the part-time private sector specialists brought in the necessary expertise and as needs for IT-expertise would shift, new contractors could be hired. Private sector IT-support contractors have often been used also at embassies.

to perform tech-support operations abroad.[28] Finally, as of 2004, the electronic infrastructure of MFASR had been connected to the electronic networks of the EU—COREU/COURTESY, ESDP-NET and EXTRANET, which carry classified traffic (mostly at 'EU-Restricted' level) pertaining to CFSP and ESDP between EU-institutions in Brussels and governmental agencies of EU-member states.

6.4 *IT-effects on Hierarchy*

6.4.1 *Challenges to bureaucratic hierarchy*

Although one could find in 2003 that e-mail has become the most common form of communication throughout the Slovak foreign service and documents were commonly exchanged in electronic form, the procedures, rules, and guidelines for document handling and filing remained paper-based. Organizational units within headquarters had shared electronic databases where all the employees of a particular unit can access electronic files of other colleagues in that unit (as long as documents are actually saved in the folders accessible by colleagues). There was, however, *no shared electronic archive* where one could look for documents produced by other units in the organization. As the 1998 User Manual for the Work in the Information System of MFA SR[29] states:

> Every user has access to his/her personal files and to other files to which s/he is granted access (mainly shared agenda in his/her particular department). Other data (agenda of other departments and users) are for him/her invisible and inaccessible (p. 2).

Every unit was responsible for keeping records and documents for the last 2 years (the so called *operative documents*). After 2 years, documents were moved to the central archive, where the staff according to an

[28] Before the global on-line network was in place, approximately 80% of the annual budget of DISC would be used to cover expenses related to travel of its specialists fixing technical problems at embassies. Since the global online network was launched, the situation has reversed and 80% of DISC budget can now be used for hardware and software acquisitions, training and outsourcing.

[29] *Užívateľská príručka pre prácu v informačnom systéme MZV SR* [User Manual for Working in the Information System of MFA SR], Department of Information Systems, MFA SR, September 1998.

elaborate categorization system decides what is to be kept in the archive and where it is to be discarded. In other words, the most recent files (always from the last 2 years) were located in a fragmented manner on individual PCs at organizational units throughout the headquarters and at embassies, and foreign service officers in need of a particular document needed to look for them by asking around the organization.[30] The central registry of all documents was in the Archive, but that was located in a separate building adjacent to the main building of the headquarters, so when officers needed to look for documents in the central archive, they had to physically cross the yard themselves or have archive employees deliver documents to them. One of the secretaries at the Archive Bureau found it particularly annoying having to move back and forth between the two buildings several times a day carrying archive documents. As she pointed out, she usually would not bring up every document right away, but would wait till she has several requests and then collects several documents to be carried over to the main building.[31] The time lags that hence ensued are obvious.

The Ministry considered the establishment of an electronic archive and there were several software companies offering comprehensive solutions, but this process was stalled in 2003.[32] This was partly due to lack of financial resources for implementing such a process, but first and foremost because there were no government-wide standards and guidelines for electronic management of documents. Another factor slowing the process down was the apparent lack of interest among the diplomats in establishing such a system. As one of the respondents observed in 2003:

> There is a mentality of 'every man for himself' and therefore every unit has its own set of documents, its own little archive. They are simply not interested in having a shared electronic archive.

Until a system for electronic handling of documents (including a shared electronic archive) is introduced, the institutional memory will remain *fragmented* in particular organizational units, and/or only *slowly accessible* given the paper-based archive routines and the physical restrictions this situation brings about.[33]

---

[30] Based on 2003 interviews.

[31] Based on 2003 interviews.

[32] Based on 2003 interviews.

[33] Establishment of an electronic archive was considered as part of a broader organization reform process initiated in 2007.

This has clear implications for possibilities of electronic cooperative efforts across formal boundaries of organizational units. Given the paper-based rules and routines and the non-existence of an electronic archive, such cooperation happened only on an informal basis. If an officer did not have any informal relations with officers in a unit where a particular document or set of documents were stored, s/he had to follow the traditional bureaucratic procedure, write a request for documents and send it through a standard procedure following the lines of authority. One of the informants described the processes as follows:

> If I know someone in a particular unit, it is no problem to send him or her an e-mail asking for information or documents that I need. However, if I do not know anyone there, I need to submit a formal written request for information to my director, who then has to submit the request to the respective head of the unit from which I want to receive documents or information.[34]

This traditional bureaucratic procedure often caused severe delays in the production of outputs.[35] What is more, in the absence of an electronic archive or at least an electronic database containing a collection of informational resources on central issues on the Ministry's agenda, junior desk-officers' requests for information from other organizational units often lacked specificity, so that documents they received from other units did not necessarily respond to their specific task-related information-needs. As a result, as pointed out in the January 2003 work-in-progress version of an internal document analyzing the possibilities for organizational improvements,[36]

> such general requests for information [...] only lead to an overload of information and documents, which in turn leads to production of outputs with low specificity and low usability of those outputs by the Minister or the state secretaries (p. 13).

An interesting observation made by one of the respondents in 2003 is that it is the junior officers who have been most pronounced in demanding the establishment of at least some form of an organization-wide

---

[34] Based on 2003 interviews.

[35] Based on 1999 and 2003 interviews.

[36] *Analýza východísk pre optimalizáciu fungovania MZV SR a rozpracovanie niektorých konkrétnych opatrení* [Analysis of Possibilities of Organizational Improvement at MFA SR and Suggestion of Some Specific Measures], internal document, Bratislava: MFA SR, work-in-progress version, January 2003.

electronic archive. The reason for this was apparently that lacking the necessary breadth of informal contacts upon their arrival to the organization, junior officers had a hard time figuring out where in the organization documents and other informational resources might be located, or whom to ask for information on particular issues they worked on. Senior officers, who had been in the organization for some time, did not have such a problem, because, for one, they knew approximately who has been previously working with what agenda, and secondly, they often also knew the right people so that they could get access to documents when such a need would arise. Hence, the absence of an electronic archive strengthened the senior officers' position as informational gate-keepers in relation to junior officers, which in turn strengthened hierarchical relations.

In addition to the formal procedural factors, factors such as seniority also played a role in terms of possibilities for effective exchange of documents across organizational units. Without previous informal contacts, junior officers usually had a hard time getting appropriate feedback and documentation from senior officers in other organizational units. Such cultural factors hence also hamper effective sharing of information throughout the headquarters despite available electronic connectivity.

The Ministry has been aware of the inflexible nature of its bureaucratic procedures. As stated in the above mentioned internal analysis of the potential for organizational improvements,[37]

> [f]rom an organizational perspective, the internal structure of the MFA is today based on vertical relations, connected at the level of super-ordinate units and the respective senior managers (i.e. at the level of department directors within sections, at the level of directors general within the Ministry as a whole). When it comes to operative effectiveness, this structure is fairly cumbersome, putting heavy pressure on the steering levels which has negative effects upon the ability of the organization to fulfil its tasks as quickly as necessary. As the experience of other MFAs suggests, the slow downs can be avoided by implementing some form of *horizontal networking* at the operative (executive) level in the preparation of documents, while the vertical line is involved first at the level of capping off documents. This pre-supposes an increase in the responsibility at the desk-officer level and a continuous increase in the skills and expertise of the staff at this level. The goal should be the establishment of a *matrix model* of organization at the MFA, in the sense of a harmonization and

[37] *Analýza východísk...*

> an increased connectivity of its units both at the vertical and at the horizontal level (pp. 3–4, my translation, italics added)

As a follow-up on this report, an internal report entitled *A Modern Ministry of Foreign Affairs* was produced. While the draft of this document was discussed at an internal meeting of the directors general and selected department-directors in early 2003, it did not bring about any specific organizational changes. In fact, even officers directly involved in the work on the document, were skeptical as for the potential for the suggested changes in structure and procedures to be implemented.[38]

6.4.2 *Challenges to the hierarchy between headquarters and missions abroad*

Dependence on paper-based standards for document-handling, non-existence of a foreign-service wide electronic archive, and standard operating procedures for communication between posts abroad and headquarters (see Figure 5 below for a model of communication between headquarters and missions abroad) limited the potential of IT-tools in place for greater integration of informational resources and shared work on tasks that the global online network connecting embassies and headquarters offered. As late as 2007, communications from missions abroad had the traditional standardized paper-based formats such as *telegram in claris* (*TIC*) or 'record of a conversation' (*záznam o rozhovore* or *ZOR*). Nevertheless, compared to the situation around 1998, the communication processes were faster and less complex in the first years of the 21st century. This was mainly due to the streamlining of software throughout the entire foreign service which at the time of fieldwork used more or less the same software (previously documents produced using various software had to be converted into formats used by recipients) and also due to the fact that documents could easily be saved by the staff at headquarters

[38] During my research stay at the Department of Analysis and Planning in May 2003, I received a work-in-progress version of the report (to my knowledge there never was a final version). It included an analysis of the implications of the changing foreign policy environment for the conduct of diplomacy and suggestions of possible pro-active organizational strategies to negotiate the newly evolved conditions. After "working the corridors" for some days trying to find out who was the author of the think-piece, all the threads eventually led to a junior officer on the staff of the Head of Administration of the MFA. Talking to the officer, who allegedly wrote the think-piece, I was struck by the high level of scepticism towards the practical use of the ideas featured there and by a good deal of sarcastic comments targeting the fact that someone like me could actually take those ideas seriously.

Figure 5: Position of the Department for Informatics and Special Communication (DISC) in the communication flows in the Slovak Foreign Service. (Based on 2003 interviews with MFASR staff.)

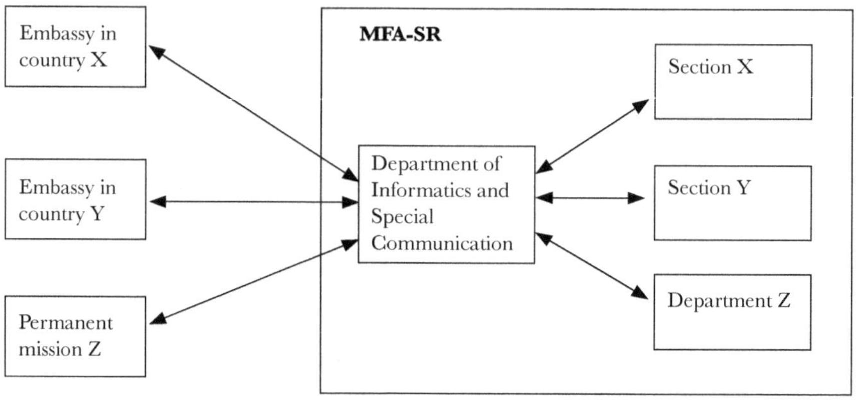

into the electronic tray of DISC. Nevertheless, before sending out a document, DISC personnel still need to see the document signed on paper. In cases when documents needed to be communicated quickly, administrative personnel from units at headquarters were forced to physically deliver documents up to the 5th floor to the DISC offices.

Further procedural inefficiencies arose due to the centralized communications-flow. Secretaries usually picked up incoming communications from embassies in the respective sectional in-tray at DISC about twice a day—early in the morning and early in the afternoon. This meant that communication from embassies received at DISC during the day often reached the respective directors at HQ in late afternoon and was hence often not dealt with until the next day. This led staff at embassies to send documents twice—they sent it to DISC as the formal procedure requires, but e-mailed or faxed also directly to the respective recipient section, so that the secretary would get it on her table right away and the document would not have to lay around in the sectional in-tray until being picked up later in the afternoon. E-mail communications should theoretically have been a solution to the problem, as every MFA SR employee had had a personal e-mail account since early 1999. However, despite the availability of e-mail, when a head of an embassy wanted to be sure that a particular director at HQ received a particular communication as quickly as possible, the safest bet still was the fax. The reason is quite simply the fact that many senior officials at the Ministry were not sufficiently IT-savvy or worked using established paper-based routines. As one informant put it in 2003:

> If you want to be sure that a director at headquarters reads your communication as soon as possible, you shouldn't use e-mail, because some directors check it only once a day, some only once a week and there are even some who only check it once a month. A direct fax to the secretary [of the director] usually gets on the director's table quicker.

Clearly, in such a situation, one rather resorts to older communication technologies such as fax, to the handling of which there are established and well operating routines.

Besides these procedural limitations, a number of the senior diplomats occupying most of the leadership positions in the Ministry are *apriori* skeptical of any increased integration between HQ and embassies. As one director general argued:

> The Headquarters decide on foreign policy, the embassies provide information, data and their points of view. The steering role of the Headquarters has to be maintained! There is no other way!

This way of thinking about the division of labour between embassies and HQ leads embassies to produce unnecessarily high numbers of reports often with low analytical value. As a number of senior respondents observed, they find it hard to sort through the electronic information overload produced by embassies. Moreover, possibly as a means to generate a semblance of productivity and importance of one's work, many embassies send in classified and encrypted reports, which need to be handled according to special procedures and about 40 officers at HQ need to read them although their contents do not necessarily warrant the high levels of encryption. Similarly, as in a number of other foreign services (Bátora and Neumann 2002, Ross 2007), assigning documents a high classification level is a way to capture attention for one's work in the Slovak foreign service.

MFASR was also considering to centralize at headquarters the performance of particular administrative functions performed at embassies. In an internal memo, the Belgian foreign ministry with accounting centralized at headquarters was mentioned as a possible model for MFASR.[39] Given the paper-based standards of communication between embassies and HQ and the non-existence of a shared global electronic archive, the hierarchy between headquarters and missions abroad had so far not been challenged to any considerable degree.

---

[39] *Analýza východísk*..., p. 13.

6.5 *IT-effects on Secrecy*

Although secrecy is an important norm in the Slovak foreign service, it is hard to speak of a distinguished culture of secrecy at MFA SR. As one senior career diplomat with background in the Czechoslovak foreign service argued,

> in this Ministry there is too much sharing of information going on. When documents are sent, too many organizational units are copied. Too many people know too much about too many things.[40]

The process of Slovakia's integration into EU and NATO has brought about an extensive review of information security procedures at the MFA SR. Several inspections of the Ministry were performed by NATO expert teams in the pre-accession period. As part of the standardization on information security standards, the NATO 4-level classification system was adopted (restricted, confidential, secret and top-secret).[41] In addition to this change in formal regulations, a comprehensive review of security clearances of all foreign service personnel has been going on. These clearances are authorized by the National Security Authority. A problem that has been particularly pronounced during the 2003 phase of my data-collection, was the fact that a number of senior diplomats (most with background from the Moscow State Institute of International Relations) did not obtain security clearances. Operations of the Ministry were hence severely hampered due to the fact that many of its senior officers were not allowed to access information at the secret or top secret level. In such a situation, special procedures had to be implemented for communication of classified information between HQ and particular missions abroad, such as for instance the permanent delegation to NATO. In the absence of a secure on-line connection between HQ and this mission,[42] all classified documents had to be communicated through cryptofax via the National Security Authority in Bratislava, who then distributed them further to the respective recipients in the capital (usually the MFASR or the Ministry of Defense) or vice-versa to the mission. This had implications for operational effectiveness of

[40] Based on 1998/99 interviews.

[41] Act 215 on the Protection of classified information and on the amendment and supplementing of certain acts, Slovak National Council, 11 March 2004.

[42] According to interview information, its non-existence was also due to delay in receiving authorization from NATO to establish such a connection.

the mission to NATO. In crisis situations, where fast reactions were required, it was virtually impossible for the Slovak diplomats in the permanent delegation to NATO to receive operative instructions from the headquarters quickly, i.e. in one or two hours.

6.5.1 *Approach to electronic risks*

As pointed out above, headquarters and missions abroad are connected by two networks, from which one is dedicated to document production and Internet access and the second one serves for classified communication. The latter network is not connected to the Internet at any point. The information security experts at DISC register and intercept approximately one attempt per month by external actors to crack into the system of the Ministry.[43]

On the unclassified network, e-mail and Internet connections have been available from virtually all workstations at headquarters since 1999. E-mail is readily used for information exchange inside the Ministry, with embassies (parallel to the often cumbersome paper-based procedures) and with external actors cooperating with MFASR. This indicates that the Ministry has moved away from the risk avoidance paradigm that characterized its approach to Internet communication prior to 1999, and towards *risk management*, which makes MFASR similar to N-MFA and DFAIT.

In this context it was interesting to note the somewhat unusual practice at a number of embassies of using free e-mail providers such as Yahoo, Hotmail or Wanadoo for their official e-mail services, which they also publish on their official web-sites as official e-mail addresses of the respective mission.[44] A number of Slovak ambassadors and lower ranking diplomats use these free e-mail accounts also for their own personal communication while stationed abroad, which though they often also use for official communication. Need for quick communications here seems to have prevailed over risk avoidance.

Finally, prior to 1999, hardware acquisitions were considered as representing a degree of security risk. The technical support staff of the Ministry preferred to upgrade PCs in use rather than buy new PCs

---

[43] Based on 2003 interviews.

[44] In late 2004, this included for instance the permanent mission to the Council of Europe in Strasburg (Wanadoo), the embassy in Abuja/Nigeria (Yahoo) and the embassy in Algiers (Centrum.sk).

because this was considered to be more secure. As one of the respondents put it, "you never know what technical elements the producer might put into the product".[45] This strategy was abandoned because PCs upgraded in this manner often malfunctioned, and there were problems with compatibility. The new strategy of hardware acquisitions after 2000 was to equip the entire organization with a standard type of commercially available PCs and off-the-shelf software available on the market. Apparently, risks related to hardware acquisitions are no longer considered to be as grave as previously. As one of the respondents put it:

> With PC upgrading the risks of losing important information due to technical failure were much greater than with standardized systems we now have in place.

6.5.2 *Policy of internal access to unclassified information*

Given that MFASR did not have a central electronic archive and paper-based procedures were followed, access to documents and information was determined and limited by administrative affiliation of officers at the time of fieldwork. The *need-to-know principle* remained central to the policy regulating internal access to both paper-based and electronic files. This might, however, not be the result of intentional internal information policing, but rather due to simple unavailability of technological solutions providing electronic archive access. In 2002 and 2003, solutions for introducing an electronic archive with open organization-wide access to a pool of unclassified documents were being discussed internally at MFASR. As stated in an internal memo:

> The MFA differs from other ministries by having a rotational workforce. Given these conditions, it is of key importance for ensuring the consistency and continuity of operations to strengthen the so called 'institutional memory'. Considering among other things the three relocations of headquarters and the ensuing problems in access to documents produced in previous years, it would be appropriate to establish an electronic archive of the MFA, to which all employees would have on-line access (possibly according to access rights). Among the basic features should be theme-based categorization of information and the possibility to use keywords in searches. [...] A vast majority of files can now be administered and reviewed in the electronic form, with no need for standardized formats,

---

[45] Based on 1998/99 interviews.

> because the e-mail sender is clearly identified and in most cases, in particular at the lower levels of decision-making, it is not necessary to archive the signature.[46]

A further internal memo was worked out 2003 in cooperation between the Archive Department of MFASR and DISC outlining a system for electronic management of documents, which was based on similar principles as the InfoBank system considered for use at DFAIT. The basic rationale was, as one director put it in 2003, that "almost no papers should be flying around this ministry in a year or two."[47] Early in 2008, such a system still was not in place at MFASR.

6.6 *IT-effects on Communication with the Public*

Matters of foreign policy had traditionally not been a highly debated topic in public debates in Slovakia. To raise public awareness of foreign policy priorities in the wake of Slovakia's entry into NATO and the EU, MFASR had launched a number of domestic information-campaigns between 2000 and 2003. Several units of MFASR became parts of interagency coordination teams featuring representatives from other governmental ministries, the NGO community and academia, and hammered out strategies of public information provision and public debates on NATO. Numerous NGOs, media houses, artists and academics were directly cooperating in public outreach projects with the MFA where channels of communication included the media, public discussions, lectures, seminars, student competitions, conferences and web-sites. As the then state secretary Korčok pointed out,

> The close cooperation with NGOs has been the main guiding philosophy of the MFA in implementing the Public Information Strategy. In 2000 and 2001, we have had an unprecedented cooperation with more than 50 NGOs that carried out over 70 different projects around the whole country. Their activities have been crucial to our common success in moulding public attitudes. The value of NGOs' projects lies not only in informing and discussing, but also in activating the civil society around the NATO topic which might have otherwise remained distant for a majority of our citizens.

---

[46] *Analýza východísk…*, p. 13.
[47] Based on 2003 interviews.

> It has been a great and challenging endeavor. I have also been taking part in many public events, talking to students, women, teachers, to the people on the street. It made me realize how important such discussions are. Sometimes, we for whom the security issues are everyday business, tend to forget that what is crystal clear to us might not be so self-evident to our citizens. Therefore, the communication with and among citizens about security has to remain constant. It is a great advantage of Slovakia that its NGO sector understands that, and is willing and able to contribute to, and facilitate such a public discussion.[48]

In a somewhat different approach, MFASR engaged in coordinating public discussions on Slovakia's membership in the EU between 2002 and 2003 in the framework of the so called *National Convention on the European Future of Slovakia*. It was organized as sessions of a group of selected societal actors including representatives of the NGOs, academia, religious groups, political parties and businesses, and coincided with meetings of the European Constitutional Convention. Following the finalization of the work performed by the European Convention, the sessions of the National Convention featuring experts from various societal spheres organized in working groups have been continuing under the auspices of the Slovak Foreign Policy Association—an NGO with close ties to the MFASR.[49]

While MFASR's reliance on NGOs had proven an effective way of reaching out to the Slovak public on particular issues, these cooperation patterns also entailed that the Ministry as such did not need to get directly involved in public debates. This factor combined with relatively low Internet-usage throughout the Slovak society,[50] with the consequently limited public demand for online-information, had had implications for the MFASR approach to Internet-based public communication. The official MFASR web-site was launched as late as 1998, contained limited information and was updated only irregularly.[51] After 2000, radical improvements to the MFA SR web-site were

---

[48] See *Slovakia's Road to NATO. A Story of Public Opinion* (available online at http://www.foreign.gov.sk/En/add.php3?text=Slovakia and NATO&file=file589.shtml)

[49] See http://sfpa.sk/en/index.php?tema=national%20convent&page=tema

[50] Only 22.1% of respondents to a survey conducted in Slovakia in 2005 answered they used the Internet "regularly". As many as 50.6% answered they "never" used the Internet. See Velšic (2005:9).

[51] An illustration of the ignorance towards updating the web-site could for instance be the fact, that even though there was a change in leadership of MFASR (a new Minister, State Secretaries and Head of Internal Administration, and a number of other personnel changes) in October 1998, the web-site was not updated on the changes

made, however. About 12-times more news- and press-releases were being posted in 2003 than in 1999 and audio-files from press-conferences were included. The web-site also featured a full-text archive of treaties between Slovakia and other countries, travel advisories and a host of other features. In 2005 the web-site was providing an extensive informational resource base on Slovak foreign policy and on activities of the Ministry.[52]

6.6.1 *Direction of communication*

In terms of other possibilities for citizens to communicate with the Ministry electronically, the web-site provided descriptions of the administrative units of the Ministry including telephone numbers to the minister's secretariat, to the state-secretaries and to the section- and department directors. No e-mail contacts were provided. Electronic requests and citizens' feedback can be sent in through an electronic form featured in the section of the site informing the citizens about the Freedom of Information Act. Yet, given the paper-based practices still in operation at the time of fieldwork, the policy for responding to citizens' e-mails was similar to that of responding to ordinary paper-letters. In 2003, there was one employee at the Press and Communications Department responsible for answering such enquiries. The employee had to print all incoming e-mails on paper and assign them a registration number by registering them on paper in a registration book, where he had to register the name of the citizen, his/her e-mail address, the topic of the query and the date when the query was received. This procedure alone indicates that as late as 2003, there were not many electronic inquiries directed at the foreign ministry by the citizens, which might have to do with the above mentioned relatively low level of Internet penetration in Slovakia. As one of the interviewees pointed out, the bottom-line rule for dealing with citizens' enquiries is that a "we-don't-know reaction is not to be used. We should at least be able to recommend the citizen where to turn to find the required information".[53] Nevertheless, there seemed to be a level of skepticism within MFASR to general citizens' enquiries. An example, was a young foreign service

---

until February 1999. In the meantime it displayed the curricula vitae, photos and other information on the previous leadership.

[52] See http://www.foreign.gov.sk

[53] Based on 2003 interviews.

officers in the Press and Communications Department (not featured in the list of interviewees), who found it very amusing during a lunch-time conversation that some small business owner from Eastern Slovakia had written an e-mail to the Ministry's common e-mail address enquiring about the possibilities for exporting slippers to the Ukrainian market. To this officer such a request seemed utterly unreasonable because the Ministry "has no capacity to deal with such trivial issues". The elitist attitude of the Ministry in relation to the society at large surfaces in such comments.

In addition to the one-way communication through press-releases and other postings, the web-site did feature a section for the so called *Inter-ministerial review process*[54] with an open full-text database of drafts of legal proposals and other policy documents submitted to the government for approval. The general public have had the opportunity to read-through and comment on legal-proposals and reports—every document posted featured the e-mail address of the officer in charge at MFASR and the deadlines for when comments and suggestions are to be received. The procedure was, however, directed primarily at other agencies of the Slovak government, which are hence provided the possibility to have their concerns included in draft-proposals of legislation. No specific procedures had been developed at the MFASR for handling eventual inputs from the general public.[55]

Finally, An innovative IT-supported initiative involving a direct consultation of foreign public actors was foreign minister Kukan's video-conference with members of the Danish parliament and Danish media in April 2003. A video-conference room was set up at the MFASR and a Danish private company was contracted by the Slovak embassy in Copenhagen to provide compatible video-conference support in the Danish capital. The Danish representatives were then summoned to the office of this private company and discussed Slovak foreign policies, issues related to the upcoming Slovak membership of the EU and NATO with minister Kukan. MFASR had attempted to organize a similar video-conferencing event in London, but the technical equipment in the House of Commons that was to be used was outmoded and incompatible with the systems used by MFA SR, and the video-conference hence had to be called off on short notice. While

[54] "Medzirezortné pripomienkové konanie"

[55] Based on 2003 interviews.

such IT-supported events have fostered a form of consultation between MFA SR with foreign publics, they should rather be considered PR-stunts and not regular policy consultations.

The policy of extensive openness that the MFASR has been practicing using its web-site has increased transparency about the foreign policy activities of the government radically when compared with the situation in 1999. Nevertheless, the publication of documents and information on the web-site serves the purpose of informing the public about foreign policy activities, but there were no possibilities for the public to get involved in consultation of foreign policy. Hence, similarly to the practices at N-MFA, web-sites of the MFASR by and large communicated information to the citizens in a one-way fashion.

6.6.2 *Level of centralization of public communication*

Although there were no formal guidelines at MFASR in 2003 as to who was and was not entitled to communicate with the media, informal practice had been that the MFASR had been communicating with the general public and the media through the highest political echelons (the minister or state secretaries) or, most often, through the spokesperson and only in exceptional circumstances through the head of the Press Department and individual ambassadors/officers.

This centralizing view had been pervasive also in the development of procedures for the management of embassy web-sites. Until early 2003, development and maintenance mission web-sites used to be the responsibility of the respective mission. Depending on the IT-skills of mission staff or on the ability to have web-sites constructed and maintained by local web-designers on a contractual basis, a few embassies had web-sites, while most did not until 2003. In 1998, there were hence only 3 operating embassy web-sites, each with a differing design and informational content.[56] As of the summer 2003, a new strategic approach was introduced by the Ministry in which all embassy web-sites were located at a central server of the Ministry and they had a standardized design and contents. On the MFA SR web-site there was an embassy web-page[57] including an alphabetical search mechanism for

[56] Embassies in Ottawa, Tokyo and Washington.
[57] http://www.mfa.sk/zu/index/index.php?lang=en

the 86 web-sites of Slovak missions abroad.[58] As the mission web-sites were located at a central server at the headquarters, their maintenance and updating was also done centrally. The problem is, though, that there was only one officer in the Press and Communications Department at HQ responsible for updating all 86 mission web-sites in 2003. As this officer also had other responsibilities, there were often delays in how often web-sites were actually updated with new information. In such a situation, some embassies (e.g. Washington) chose to have another web-site besides the official web-site at the Ministry server. In such cases, the official web-site at the Ministry only contained general information (e.g. contact information, CV of the ambassador) which did not need to be updated very often, while the web-site maintained by the embassy contained information on upcoming cultural events, current travel advisory, embassy statements on current political events in the country of its location etc. Given the fact that most embassies used the standardized web-site format maintained by the Ministry, such a double presence on the web of some embassies might lead to uncertainty as to what are the official embassy web-sites in those cases and where to look for updated information.

6.7 *Conclusion*

On par with other governmental agencies, MFASR had radically modernized its IT-infrastructure between 1999 and 2004. Internet access was enabled from virtually all workstations, employees were provided with personal e-mail accounts and online network connections were established between headquarters in Bratislava and missions abroad. By and large due to the non-availability of government-wide standards for electronic management of documents, these technological improvements at MFASR have been carried out without corresponding update of organizational procedures, routines and processes, which remained largely paper-based. In such an intra-organizational environment, boundaries between official jurisdictional areas of organizational units were strictly maintained and the line hierarchy was strictly followed.

[58] In late 2004 this included web-sites of 58 embassies, 1 embassy branch office (in Bonn/Germany), 8 permanent missions to international organizations, 1 Economic and Cultural Institute in Taipei/Taiwan, and 8 Slovak Cultural Institutes.

Organizational units remained formally linked at the level of directors with horizontal electronic exchange of files and documents between desk officers and operative staff mostly conditioned by informal relations. E-mail and other means of electronic connectivity, which are readily available throughout the organization hence do not challenge the established Weberian system of bureaucratic hierarchy.

Paper-based communication standards were also limiting the way IT could be used in the communication between missions abroad and headquarters. All official communication flows through a central point at the headquarters. Given the lack of connectivity and non-existence of a global electronic archive, the hierarchy between headquarters and embassies is maintained. These findings suggest that effects of IT on hierarchy at MFASR match the *status quo scenario*.

The technological solutions in place allowing for Internet access and e-mail communications for all employees suggest that in terms of the approach to electronic risks, risk avoidance that characterized most of the 1990s was replaced by *risk management* at MFASR. Moreover, due to paper-based standards and non-availability of a shared electronic archive, the internal policy of access to unclassified information had been informed by the need-to-know principle. In all, these findings suggest a match with the *status-quo scenario*.

Compared with the situation in 1998–99, positive changes have been recorded in the Ministry's use of its web-site in publishing information. In 2004, the web-site was far more comprehensive and included extensive archives of documents, speeches and other information on Slovak foreign policy and on activities of the MFA SR. Embassy websites were integrated into the main site, located at a central server in the headquarters and received a standardized design. Despite these positive tendencies in using the web-technology, MFA SR activities on the web remain at the level of informing the public about foreign policy and activities of the Ministry in a one-way fashion. Possibilities for feed-back are limited, and the Ministry does not encourage the public to engage in consultation on foreign policy issues. Moreover, the channels of MFASR's communication with the public are centralized and individual units do not communicate with the public. These findings match the *status-quo scenario*.

As a new foreign ministry of a new state, MFASR had been operating in the state of a permanent performance crisis. The crisis had been related to the factors causing performance crises to other foreign ministries including the information-intensive environment brought about by the IT-revolution and other impacts of globalization including the rise of non-state actors. As was shown in the cases of Norway and Canada, facing uncertainty, foreign ministries were turning to exploitation of established routines and procedures, which often led to limits in using IT. What made the overall performance crisis all the more profound in the Slovak case was the fact that MFASR had been operating in a domestic environment, where basic axioms of the political system were being defined and basic features, legal rules and procedures of public administration were being developed and standardized. This led for instance to non-availability of standards for electronic management of official files with the consequence of paper-based standards being applied throughout the Slovak government despite the broad availability of advanced IT-tools. Hence, while the findings from MFASR match the status-quo scenario, this can hardly be related to a rigidity of the organizational system in place.

At least four factors might contribute to rather fast implementation of new IT-based procedures and working processes at MFASR. Firstly, following Slovakia's accession to the EU in 2004, the basic legal framework is in place enabling the government to implement government-appropriate procedures for electronic management of documents in the public administration. Secondly, Internet-penetration in the Slovak society had been increasing rapidly in recent years and this might be leading to increasing pressures on the government to provide services in electronic form.[59]

Thirdly, in October 2005, Slovakia was voted in to become a non-permanent member of the UN Security Council. In the Fall of 2005, MFASR was in the process of establishing a global electronic data-base connecting the headquarters with embassies and providing open access to a pool of documents pertaining to the work in the UN Security Council. The effects of Slovakia's membership in the UN Security Council in 2006 and 2007 on the work processes in the MFASR would need to be more thoroughly analyzed.

Finally, as briefly mentioned above, the MFASR has been involved in a process of comprehensive organizational reform with the aim of introducing a "21st Century foreign ministry". The project was

[59] In 2004, the number of individuals in Slovakia using the Internet for obtaining information from public authorities was 20.9% (EU-25 average was 21.2%), and the number of enterprises was 42% (EU-25 average was 45%). (Based on *Eurostat*).

launched on November 23, 2006 under the acronym TREFA.[60] Private management consulting companies were hired to conduct a comprehensive organizational audit of work-flows, procedures, administrative effectiveness, relations between line officers and directors, cross-unit cooperation, use of IT in support of work processes, and to suggest possibilities of organizational improvements in all the areas concerned. This was done throughout 2007 and continued in the first months of 2008. The overall budget allotted to this program came near to 17 million SKK (more than 0.5 Million Euro). Based on the recommendations of the audit reports, the MFASR was to implement a comprehensive program of organizational change. The program had backing at the senior leadership level including the foreign minister Kubiš, but first and foremost the head of administration, Peško. Daily project management of TREFA was entrusted to a project manager reporting directly to the head of administration. Experienced diplomats were put in charge of change management task forces on a) work processes; b) foreign policy strategy; financial management; c) human resources; d) information technologies; and e) external communication. As the project manager of TREFA got an ambassadorial post staring in January 2008, a new project manager took over. Further analyses would be needed to assess the effects of TREFA in the MFASR. What seems clear at the time of writing is that the TREFA-project is the first attempt by the MFASR to think systematically through IT-supported organizational reform and assign resources to its conduct.

In the next chapter, the findings from the three case studies are summarized and conclusions are drawn.

[60] The acronym TREFA stands for "*Projekt trvalej efektívnosti riadenia*"—Project of sustainable management effectiveness. "Trefa" is also a slang word for "score" or "hit" in sports related terminology.

CHAPTER SEVEN

CONCLUSION: THE ORGANIZATIONAL BASIS OF DIPLOMACY RENEWED

7.1 *Introduction*

This study was motivated by a set of analytical lacunae surrounding the question much debated by academics and governments in recent years, namely: What (if any) are the effects of IT on the organizational basis of diplomacy? Numerous interesting and valuable observations have been made in the body of contributions addressing effects of IT on the way diplomacy is organized and conducted. Still, reviewing the literature, it was difficult to find a clear and consistent set of arguments as to what it is about the organizational basis of diplomacy that IT might be challenging. Moreover, it was not entirely clear from the available contributions what aspects of the organizational basis of diplomacy would have to change if we were to declare that the organizational basis of diplomacy had changed. In other words, although valuable and insightful, previous analyses of IT-effects on diplomacy tell us little about the core organizing principles of modern diplomacy as an institution of the state order, and indeed little about how IT relates to and possibly challenges these organizing principles.

Anchored in new institutionalist approaches in political science and organization theory, this study, firstly, developed a conceptualization of diplomacy as an institution, i.e. a set of rules, norms and routines regularizing relations within the system of states structured by the constitutive logic of sovereignty. Hierarchy, secrecy and one-way public communication were identified in Chapter 2 as the core organizing principles sustaining the 'informational prerogative' of the state in foreign affairs. These organizing principles are reflected in formal, structural elements of foreign ministries, which constitute the global organizational basis of diplomacy (see Table 1 and Figure 1 in Chapter 2 for conceptual models). Obviously, there are other features characterizing diplomacy as an institution, including, for instance, diplomatic language, the norms of extraterritoriality and diplomatic immunity, specialized diplomatic socialization procedures, the principle of staff rotation between home and abroad etc. (see for instance Berridge [1995] and Bátora [2005] for elaborations of these and other institutional features of diplomacy). But, as pointed out in chapters 2 and 3, a focus on hierarchy, secrecy and one-way public communication is of particular relevance when we study the effects of IT on diplomacy as an organized set of practices and procedures regularizing communication between states. These three core organizing principles are constitutive of modern diplomacy as an institution of the state order and a focus on them enables us to get a clearer vision of the institutional character and identity.

Chapter 3 then addressed preceding waves of innovation in communications related in particular to the introduction of the telegraph in the 1850s and the effects this had on the foreign ministries, which was followed by a discussion of the broad change dynamics in the international environment related to the ongoing IT-revolution. As a first step in the development of an analytical framework used in this study, the core organizing principles of diplomacy were then operationalized. This was followed by the development of three scenarios of possible effects of IT on these principles embedded in foreign ministries, including transformation, status-quo and renewal. The latter two represent different degrees of path dependent change in the organizational basis of diplomacy, while the former one involves radical change of the organizational basis of diplomacy.

More specifically, *transformation* represents a radical form of change that may emerge at the level of individual foreign ministries as a result of institutional performance crises, institutional collisions, skilful change-management or a combination of these. At the level of diplomacy as an institutional order, it was suggested that such a change would involve a radical change of established norms, rules, structures and procedures leading to the emergence of a new set of IT-supported institutional principles for organizing diplomacy. In this scenario, complex networks of actors replace states as primary actors in foreign affairs, which leads to new notions of who is a diplomat, what are appropriate diplomatic situations and what is appropriate diplomatic conduct, and thereby leads to change *of* diplomacy as an institution.

The other two scenarios—status-quo and renewal—involve a degree of structural adaptation at the level of individual foreign ministries, but no fundamental change of basic notions of who is a diplomat, what are appropriate diplomatic situations, and what is appropriate diplomatic conduct. *Status-quo* represents a situation where IT implementation has no or only marginal effect on established organizational structures and procedures in individual foreign ministries. In the *renewal* scenario, organizational structures and procedures of foreign ministries are changed with the help of IT and new structures and procedures are introduced. By using IT in new ways to improve internal information management and to include external informational resources, the foreign ministry's role as the guardian of the state informational prerogative in foreign affairs is maintained. Renewal hence does *not* entail new notions of who is a diplomat, what are appropriate diplomatic situations and what is appropriate diplomatic conduct. Both of the latter scenarios hence represent change *in* diplomacy as an institution, albeit different magnitude(s) of such a change.

The institutionalist approach alerts us to the possibility that the character of the state and a foreign ministry's organizational history, its role in the government and national-administrative specifics may in various ways be reflected in the way the conduct of diplomacy is organized in various states. This in turn has implications for the way IT is implemented and taken into use at any given foreign ministry. Hence, to study the effects of IT on diplomacy as an institution embedded within states and more specifically in organizational arrangements at foreign ministries, it was necessary to conduct case-studies. This strategy enabled an in-depth data-collection and analysis of processes of implementation and use of IT within specific organizational settings.

As pointed-out in chapters 2 and 3, institutional factors have implications for paths of adaptation and avenues of change in organizations. The three foreign ministries chosen for investigation in the current study were characterized by differences in their institutional settings, which had implications for their respective position and role in their national government, and indeed for their respective approaches to IT-implementation and use. Canada is a federal state with a challenged sovereignty, which left DFAIT (and its organizational predecessors) having to continuously struggle for primacy in coordinating foreign affairs with other governmental departments, provincial governments, the private sector and non-governmental organizations. DFAIT had been a vanguard among foreign ministries world-wide in applying IT in support of its operations. The DFAIT case was chosen for the current study to represent a foreign ministry where change due to effects of IT on the core organizing principles of diplomacy was expected to be the most profound and the case could hence serve as a benchmark of the magnitude of change.

The N-MFA, on the other hand, is more than a century-old foreign ministry of a unitary state. It has been characterized by a traditionalist approach to organizing diplomacy with strict internal adherence to lines of authority, a culture of secrecy, and traditionally somewhat strained relations with the egalitarian Norwegian public sceptical of its quasi-aristocratic ambience. Organizational reform initiatives at N-MFA were traditionally met with inertia and exploitation of existing routines and practices, and the implementation of IT in recent decades was no exception. In the current study, the N-MFA was hence chosen as a case of a foreign ministry hesitant in taking IT into use and the effects of IT on the core organizing principles of diplomacy were hence expected to be least profound in this case.

Finally, the MFASR is a foreign ministry of a new state. Although it did feature a core of experienced diplomatic staff from the former Czechoslovak foreign service who brought with them standard diplomatic routines and procedures, the vast majority of employees had no previous diplomatic experience and the organization as such was not characterized by the traditional diplomatic culture. Moreover, the MFASR had been establishing its basic working procedures and organizational structures amidst the set of environmental developments commonly referred to as the information revolution which was likely to lead to early adoption of IT-supported procedures and working processes within the organization. The MFASR was chosen for the current study as a case of a foreign ministry in which the core organizing principles of diplomacy were institutionalized to a low degree, and in which implementation and use of IT were to have profound effects permeating the new organizational arrangements in the very process of their institutionalization. Indices of innovative IT-supported principles and practices for organizing diplomacy were expected to be found.

Descriptions of the cases and findings on IT-effects on the core organizing principles of diplomacy were presented in detail in chapters 4, 5 and 6 respectively. This chapter first presents a comparative summary of the findings in relation to hierarchy, secrecy and one-way public communication. This is followed by an outline of other observations and findings found in all three cases. The drivers of change are then discussed in the light of the theoretical concepts developed in Chapter 3. A discussion of theoretical lessons follows. Finally, the potential of a radical transformation of diplomacy is briefly addressed along with the factors that might bring about such a development. The latter section serves also as an outline of possible future avenues of research.

7.2 *IT-effects on the Core Organizing Principles of Diplomacy: The Main Findings*

7.2.1 *IT-effects on hierarchy*

In terms of effects of IT on bureaucratic hierarchy, effects of IT were most profound in the case of DFAIT, which was a forerunner among foreign ministries world-wide in establishing a shared global communications platform for the entire foreign-service. High levels of information sharing through e-mail, shared web-based informational resources on password-protected web-sites on the Internet and intranet, and availability of electronic archives providing access to informational resources irrespective of formal divisions between organizational units and formal lines of authority, provided a basis for project- and issue-based network cooperation in virtual teams throughout the organization including staff at embassies and external actors. High levels of sharing also related to the general openness of senior managers to e-mail communication with line officers which allowed for a flattening of the formal hierarchy of offices (there were exceptions, of course). Hierarchy was reinvigorated in relation to the user rights in the e-mail communication, where following a period of e-mail overflow due to various kinds of often semi-professional broadcast e-mail, only senior managers got the right to send broadcasts. Moreover, readiness to answer e-mails was often moderated or hampered by the excessive number of e-mails in the mailboxes of senior staff, so that having 300 or more unanswered e-mails in the inbox became an informal practice among senior officers at DFAIT. This in turn led to the need to 'reinforce' the message by personally showing up at the door of a senior manager anytime swift action or reply was required.

Hierarchy in internal communication at DFAIT was also challenged by the introduction of the InfoBank system which allowed for the creation of sharing systems between units at headquarters and particular embassies, which effectively enabled issue-based informational integration and use of available expertise on a global scale. This enabled a reversal in the traditional division of labour between headquarters and embassies, where some Canadian missions in major capitals could now perform analysis and devise policy instructions, while geographical units at headquarters in Ottawa with often work merely as a relay point for the embassy's communication with actors within DFAIT and within the Canadian government and business sector more generally. However, despite such global interconnection and the radically increasing number of electronic transactions, the use of paper had not decreased. On the contrary, there was a 400% increase in paper usage between 1993 and 1999, which indicates that formal paper-based procedures are being laboriously maintained in parallel to the electronic exchange and storing of information. While this is the case, it was shown that electronic information management systems at DFAIT allow for cross-unit and cross-hierarchy integration of teams and task-forces, and thereby an alternative informational infrastructure in parallel to the formal bureaucratic structure emerges involving units at headquarters and missions abroad. Given such infrastructural and procedural integration on a global scale, as well as the established practices of sharing information among officers irrespective of their geographical locations,[1] DFAIT worked as an informationally integrated global organization at the time of writing. Based on these findings, DFAIT matches the *transformation scenario* when it comes to the effects of IT on hierarchy.

It the N-MFA, the magnitude of change due to effects of IT on bureaucratic hierarchy was less extensive than in DFAIT. In the Oslo headquarters, the introduction of Internet-access and e-mail from all workstations came comparatively late in the summer of 2000 and communicational integration of embassies into a shared global network was not fully operational until 2002. These technological developments were not supported by appropriate updates of working procedures, so that staff continued to operate largely using established paper-based routines, which led to severe inefficiencies. Once a shared system for electronic management of documents was developed (BEST, and later 'eUD 360'), it reflected principles of bureaucratic hierarchy in a fairly rigid manner. Heads of organizational units would cap off documents electronically and this information would be stored as meta-data with the respective document. Moreover, formal hierarchical roles of 'leaders' and 'desk-officers' with specific user-rights attached to each of the roles were created in the system. This would imply, for instance, that 'desk-officers' could not electronically assign tasks and access information in the system to the same extent as 'leaders'. Through such procedures, hierarchy as well as formal authority and boundaries of organizational

[1] Close to 80% of DFAIT officers reported regular sharing of files with colleagues at HQ an at missions via e-mail and shared drives in 2001. For more detailed information see chapter 4.

units were re-enacted in the electronic transactions. Nevertheless, IT (in particular e-mail and later establishment of an electronic archive) did allow for increased information exchange across formal lines of authority and across unit boundaries at the operative preparatory stages in the decision-making processes. The implementation of a global network solution allowed for increased informational integration of embassies and headquarters and to unprecedented policy-issue related patterns of information exchange between Norwegian embassies in different geographical regions. An example was the preparation in 2001 of the Norwegian USA-strategy by a virtual team featuring officers both at headquarters in Oslo and in the missions in North America. Since then, IT-enabled involvement of officers with special expertise in the work on policy-documents irrespective of their geographical location has been an increasingly standardized practice in the N-MFA. Finally, e-mail was readily used for the previously limited communication between Norwegian missions working in the same region or addressing the same policy agenda. In sum, these findings indicate that in terms of IT-effects on hierarchy, the N-MFA matches the *renewal scenario* in the analytical framework.

In MFASR, effects of IT on bureaucratic hierarchy were the least profound compared to the other two cases. This was related primarily to the fact that despite broad availability of advanced IT-solutions throughout the organization, no shared organization-wide system for electronic management of documents was introduced, and procedures for filing documents electronically and information exchange mirrored established paper-based routines. Moreover, there was no global electronic archive and the most recent files were stored on the PCs in the respective organizational units administering them. Organizational memory was hence stored in a fragmented manner throughout the MFASR fostering a situation in which fast access to information and documents from other than one's own organizational unit are highly dependent on informal contacts with the colleagues with access to such material. Obviously, this is a severe disadvantage for junior officers entering the service and/or also for senior officers rotating back to headquarters after postings abroad. Although communication between missions abroad and units at headquarters was mostly electronic, it was flowing through a centralized node at the headquarters, which kept slowing down the communication processes considerably. Informational integration between headquarters in Bratislava and Slovak missions abroad was hence limited by and large maintaining the traditional division of labour. In sum, effects of IT on hierarchy at MFASR match the *status-quo scenario*.

7.2.2 *IT-effects on secrecy*

Contrary to the considerable variation in the magnitude of change due to IT-effects on hierarchy, there was similarity in the magnitude of change at the three foreign ministries due to IT-effects on secrecy. In their fundamental approach to electronic risks, there has been a move in all three organizations from risk avoidance towards *risk management*. As pointed out in the Equipped for the Future Report (p. 23), this involves acceptance of the fact that one is unable to ensure perfect security of electronic information systems against breaches and other electronic attacks. At the same time, though, not being connected to the Internet and not having access to quick communication by e-mail is perceived as representing a far greater risk for the foreign ministry than losing a particular piece of information. DFAIT was a pioneer in adopting this approach and enabled full access to Internet and e-mail from all workstations throughout the Canadian foreign service in the mid-1990s. Successive lines of senior political leadership at DFAIT were consistent in supporting a more liberal approach to the use of IT in daily work and partly personally pioneering new IT-supported working methods.[2] MFA SR and N-MFA followed suit in 1999 and 2000 respectively. In terms of the approach to electronic risks, all three foreign ministries hence matched the *renewal scenario*.

When it comes to internal policies of access to unclassified information, there was variation in effects of IT at DFAIT and N-MFA on the one hand, and at MFASR on the other hand. As the case of DFAIT had shown, rapid introduction and spread of electronic communication throughout the organization entailed also a gradual process of 'watering down of awareness' among foreign service officers in the handling of classified documents electronically. Risks were related also to the communication of electronic communication of unclassified information, where in particular e-mail chains copied to various recipients inside and outside DFAIT could contain information on various seemingly

[2] This includes for instance using the Internet to access updated information on international crises at a time when the world wide web was still in its infancy (e.g. Deputy Minister of foreign affairs Smith in the mid-1990s) or requiring access to e-mail via wireless hand-held devices (e.g. Foreign Minister Manley in 2001).

unrelated subjects but, when aggregated and analyzed, could represent an information security risk. Similar concerns have occurred in the N-MFA. Both Ottawa and Oslo had hence introduced organization-wide electronic archives in which access was either open or regulated by an electronic version of the need-to-know principle. This entailed that informational resources were visible but access to them was in some cases limited by the organizational units that filed the documents, which represented a form of *guarded openness*. Hence both DFAIT and N-MFA match the *renewal scenario* when it comes to policies of internal access to electronic information.

At the MFASR, there was no shared organization-wide electronic archive, and electronic documents were mostly accessible through shared databases of particular organizational units, which effectively maintained the paper-based need-to-know principle despite broad availability of various IT-solutions throughout the organization. This indicates a match with the *status-quo scenario* in the analytical framework of the current study.

7.2.3 *IT-effects on communication with the public*

When it comes to the direction of communication, there was a differentiation in effects between on the one hand the N-MFA and MFASR, and on the other hand DFAIT. Both of the former two foreign ministries developed extensive web-sites rich in foreign policy information. The N-MFA launched the so called Norway-portal, into which all the embassy web-sites were integrated. The MFASR had also launched a shared web-platform for all Slovak embassy web-sites in 2005. Yet, both of the foreign ministries continued using the web-sites for providing information to the public in a one-way fashion (press-release mode) and they did not encourage the public to get involved in public discussions of foreign policy priorities. Moreover, provision of information to the media and the public was highly centralized in both MFASR and N-MFA where the most common channel of communication was the official spokesperson or the respective minister of foreign affairs. These findings indicated a match with the *status-quo scenario* in the analytical framework of the current study.

While much of DFAIT's public communication also has the form of one-way communication, the Department also introduced a number of innovative web-based practices enabling public consultation. A pioneering step in this direction was the web-site of the Dialogue on

Foreign Policy launched in 2003 as part of a nation-wide consultation of Canadian foreign policy strategy. This was followed by a more structured form of public consultation and engagement through the *eDiscussions* section of the Canadian International Policy web-site, which was supported by procedures for evaluation, feedback provision and inclusion of the suggestions made by the public in policy briefs used in policy development by the Strategic Branch, other relevant units and the senior management at DFAIT. The fact that DFAIT reflects about the public inputs on the eDiscussion web-site and posts official replies suggests a step towards *two-way communication* with the public. Albeit still limited in scope and level of institutionalization, this practice is a departure from a broadcast style one-way communication with the public usual in foreign ministries. The fact that in 2006, the team managing eDiscussions at DFAIT was invited by several foreign ministries (including the German foreign office, the N-MFA, and Turkish MFA) to make presentations of the web-tool and organizational practices associated with it, as well as the recent efforts to launch public web-discussions of various kind by the Australian DFAT, British FCO, and the Ukrainian MFA indicate an interest among foreign ministries world-wide in exploring ways of greater public engagement.

Furthermore, use of web-technology enabled individual units within DFAIT to create their own web-sites featuring different graphic design and identity than the rest of the organization and thereby led to multiplication of channels of DFAIT's public communication. This tendency was also further deepened by DFAIT's responsibility for the development of specific sections of governmental information portals as part of the Government On-Line initiative. In terms of the impacts of IT on communication with the public, DFAIT matched the *transformation scenario*.

As noted in Chapter 4, through, the actual impacts of DFAIT's public consultations on the processes of foreign policy planning and decision-making were ambiguous. The web-based consultation process represented a new set of practices only being tested more thoroughly at the time of writing and further investigations would be needed to ascertain whether these innovative uses of web-sites indeed do represent a new mode of involving the public in foreign policy decision-making or whether this is merely a way to increase the foreign ministry's visibility and legitimacy in the domestic society.

7.2.4 *Additional observations and findings*

In addition to the findings pertaining to the core organizing principles of diplomacy, several other observations common to all the three cases can be noted. They concern effects of IT on a number of complementary organizational aspects of foreign ministries, which fill in the overall picture of the effects of IT on the organizational basis of diplomacy.

First, introduction of IT-based communication and filing within all three foreign ministries investigated here led to a *severe undermining of the institutional memory*. This was primarily related to the fact that IT was implemented mostly without proper updates in routines and procedures. In all three organizations studied here, this led to ambiguities about the status of e-mail communication, uncertainty as to what versions of documents should be filed and where. These procedural uncertainties led to data-hoarding on individual PCs, which further deepened the fragmentation of the institutional memory compared to the previously well-kept paper-based archives in each of the ministries. This was perceived as particularly problematic, as institutional memory is key to the maintenance of the institutional identity of a foreign ministry, and indeed also central to the maintenance of political accountability. MFASR, of course, is a bit of a special case, because it did not have much institutional memory to work with before 1993, but it did experience similar practices of data-hoarding and low levels of filed material as N-MFA and DFAIT. It seems hence fair to suggest that the years 1995–2005 are likely to represent what a senior Norwegian diplomat has referred to as a *'black-out decade' in foreign ministries' record-keeping*. Hence, Lord Hammond's mid-19th Century reasoning about the telegraph undermining accountability and leaving unsatisfactory records (see Chapter 3) was once again echoed in foreign ministries in relation to the introduction of electronic communication a century and a half later. Moreover, the rapid introduction of IT is a fundamental challenge to one of the key traditional roles of foreign ministries discussed by Hocking (2007:14), namely as the locus of institutional memory on the foreign affairs of a state (see also Kurbalija 2007:310).

Just as the effects of IT were similar in terms of undermining the institutional memory, so were also the three ministries' respective responses to the unstructured implementation and uses of IT. Integrated systems for electronic management of files and official electronic communication were being introduced almost simultaneously at DFAIT (*InfoBank*) and at N-MFA (*BEST* and *eUD Public 360* respectively) around 2003.

MFASR was also recognizing the need to introduce a similar integrated system and meanwhile operated using electronic versions of standard bureaucratic forms for document-exchange. The new systems for electronic management of files and information provided possibilities for *formalizing innovative connectivity across formal bureaucratic divisions and lines of authority*, in particular at the operative level. DFAIT's InfoBank project, for instance, provided possibilities for creating sharing systems and *shared databases between units working on the same agenda irrespective of geographic location*, i.e. including embassies. The N-MFA was establishing a foreign-service-wide electronic archive providing open access to a pool of unclassified documents. Such solutions enabled the work of project-teams and task-forces.

This in turn opened the possibility of increased integration of missions abroad and headquarters. As the case of the Canadian embassy in Tokyo indicates, electronic networks enable *missions abroad to take on responsibilities for policy development, strategic analysis and decisions*, while the role of geographic units at headquarters as mediators or contact points in relation to the domestic political system increases. The effect of IT is that in some cases (e.g. when particular expertise is located in an embassy) it might undermine the traditionally hierarchical relationship and division of labour between headquarters and missions abroad.

Effects of IT have also led to a degree of *confluence of working styles between missions and headquarters*. In simplified terms, work performed at missions had traditionally been characterized by spending considerable amounts of time in the field outside of the office and often regardless of actual office hours. On the contrary, work at headquarters had been performed in the office and with strict adherence to working hours. In general, the work of foreign service officers both at headquarters and in missions is now increasingly conducted by sitting in front of a computer. With mobile solutions such as laptops and hand-held devices, the practice of working irrespective of location is no longer the usual norm only in missions, but has been becoming an increasingly widespread practice among officers stationed at headquarters (in particular the headquarters in Oslo and Ottawa are examples of this). This leads to challenges related to difficulties in drawing the line between private life and work, but that is a usual situation for officers stationed in missions abroad.

A general trend related to application of secrecy in the information-intensive foreign policy environment has been the increasing speed of turning classified information into public information. As a senior officer at the N-MFA put it, "what is classified in the morning might be unclassified in the afternoon." While hence access to classified information remains a source of status for foreign service officers, it is now additionally also the *speed of access* that *provides status*—knowing something that others do not know *yet*. As a consequence, given the often cumbersome procedures for handling classified information and files which decrease the speed of information exchange, a general trend at DFAIT and N-MFA has been that of decreased use of classification for the sake of quicker communication. Again, MFASR is a somewhat special case, where the classification awareness throughout the organization has been increasing due to the implementation in 2003 of a standardized classification system used by NATO.

In terms of using the Internet in support of public communication, at all the three foreign ministries *web-sites have enabled an unprecedented level of public accessibility of information on foreign policy and on the activities of the foreign ministry*, including full-text archives of treaties, travel-advisories, organizational information on the foreign ministry, and other features. Compared with the mid-1990s, this represents significant enhancements in public accessibility of foreign policy—and consular information. All three countries had web-sites developed for their missions abroad, which hence enabled them to establish innovative *virtual representation abroad*. In the case of N-MFA, this involved for instance the possibility to have a look at what is going on in Norway through a set of web-cams placed around the country and accessible through the embassy web sites.

Table 7 provides an overview of the findings.

7.3 *What were the Drivers of Change?*

Why was the change more extensive in DFAIT than in the other two foreign ministries? As pointed out in chapter 3, in and of itself IT does not have the potential to bring about change in organizations and institutional orders. Organizational and institutional forces are necessary as reinforcing, slowing down or catalyzing IT-driven or IT-enabled change. All the three factors driving IT-enabled institutional transformation elaborated in Chapter 3 were present in various degrees and forms in the three cases investigated in this study.

Table 7: Effects of IT in the foreign ministries of Canada, Norway and Slovakia

MFA *Core organizing principle of diplomacy*	DFAIT	N-MFA	MFASR
HIERARCHY	Transformation	Renewal	Status-quo
SECRECY	Renewal	Renewal	Renewal/Status-quo
ONE-WAY PUBLIC COMMUNICATION	Transformation	Status-quo	Status-quo
Overall assessment	*Transformation/ Renewal*	*Renewal/ Status-quo*	*Status-quo/ Renewal*

7.3.1 *Sensing the danger of own irrelevance*

A sense of a severe performance crisis and an awareness of the potential danger of becoming irrelevant as the lead agency in charge of foreign affairs was a major driver of change in all the three foreign ministries. It was related first and foremost to the rapidly changing information-intensive environment generating unprecedented informational pressures on the foreign ministries. But a sense of performance crisis also stemmed from the well-established practice of continuous mutual comparison of organizational practices and performance among foreign ministries. On several occasions, interviewees I spoke to at both N-MFA and MFASR referred to DFAIT as a foreign ministry they were looking up to for innovative uses of IT in support of diplomatic operations. In addition to DFAIT, the N-MFA staff frequently also mentioned the Finnish and the Danish MFAs as the peer-ministries one looked at when studying innovative IT-usage. The MFASR, in turn also looked at the Danish MFA as a possible model of organizational modernization. DFAIT was fairly self conscious of its status as a leader in IT-use among foreign ministries, but practices in the Australian DFAT or the British FCO were occasionally also mentioned as models.

It needs to be noted that the perceptions of an imminent performance crisis did have a different character at DFAIT and N-MFA on the one hand, and in MFASR on the other hand. At DFAIT and N-MFA the sense of being outperformed by other governmental agencies, private enterprises, NGOs and other foreign ministries was clearly reflected in a number of strategic memos and reports that both organizations produced since the mid-1990s. A number of these documents framed the discussion of IT-modernization as an inevitable step that had long been taken in globally operating private enterprises or NGOs (see reports by DFAIT discussed in Chapter 4). Such a shift in the objects of benchmarking and comparison from the traditional focus on other foreign ministries towards a much more heterogeneous group of transnational corporations and globally operating NGOs is a driver of institutional change as the fundamental notions of the foreign ministry's identity are hence being extended. As Morrill (forthcoming) points out, such an 'elasticity' of reference frames may bring about a lasting form of institutional change. Yet, the traditional identity of foreign ministry was also activated when the arguments for the need of IT modernization were framed as a "matter of national interest" (The Electronic Foreign Service Report produced by the N-MFA in 1999).

At MFASR, on the other hand, performance crisis was more-or-less a standard mode of operation since it was founded in 1993. It had relocated several times between 1993 and 1998, it had to develop procedures and routines more or less from scratch, hired high numbers of non-professional diplomats, and it struggled with frequent shifts of the organizational structure and procedures.[3] Hence although all three foreign ministries were adopting IT in a somewhat unstructured way in the late 1990s, at the MFASR this was merely a part of the overall process of gradual standardization of routines, while at DFAIT and N-MFA this represented an overhaul of routines of record-keeping and filing, as well as a strain on internal communication patterns (see parts 4.4 and 5.4 above). To some extent then, the performance crisis following the introduction of Internet and e-mail access at DFAIT and N-MFA and leading to the above mentioned fragmentation of institutional memory was more severe at these two foreign ministries than at MFASR.

Responses to the performance crisis led at DFAIT and N-MFA to the development of central electronic archives, unified user-interfaces

[3] It needs to be noted, though, that the effects of this organizational 'ad-hockery' in the first fifteen years of MFASR's operation, which to some extent continued early in 2008 due to, for instance, the non-existence of a foreign service law in Slovakia, were ameliorated by the consistently high levels of acceptance and public support for the work of the MFASR among the Slovak population.

and standardized procedures for electronic management of documents. Moreover, the Canadian and the Norwegian foreign service were at the time of writing operating as global on-line network organizations connecting informational resources of headquarters and missions abroad and fostering cooperation across organizational boundaries and geographical locations, which provides the basis for the above conclusions about the renewal-effects of IT. At MFASR, which had been operating with a constant crisis of performance brought about by using IT in combination with inefficient paper-based routines and with no integration of informational resources on a global scale, but lacking a sense of performance crisis due to high levels of approval in the domestic society, the response to the actual performance crisis had been slow resulting in limited IT-supported change of the organization.

7.3.2 *IT-induced collision between secrecy and hierarchy*

As the three case studies had shown, one of the key pre-conditions for the integration and effective use of the knowledge base in the three foreign ministries was the availability of Internet-connections from all workstations and the establishment of openly accessible pools of unclassified documents. This was directly challenging the established standards of information security and established procedures ensuring a proper level of information protection within a foreign ministry (e.g. the need-to-know principle). An institutional collision ensued between, on the one hand, the need to maintain secrecy norms and procedures providing the foreign ministry with a level of discretion in decision-making and, on the other hand, the need to operate flexibly with the best possible access to information and expertise and hence retain the role as the lead agency in the conduct of foreign affairs. In all the three foreign ministries investigated here, the latter concern was eventually prioritized and information security procedures were liberalized to enable greater internal integration of informational resources and more flexible working methods (including mobile devices).

There were again nuances between on the one hand DFAIT and N-MFA and, on the other hand, MFASR. In the former two, the imminence of the need to apply IT in support of operations was in addition to internal calls for change also related to external pressure by other governmental and societal actors in Norway and Canada questioning the continued relevance and legitimacy of the foreign ministry's leading role. Introduction of IT and the ensuing institutional collisions between the principles of hierarchy and secrecy at DFAIT and N-MFA had been hence catalyzed also by heavy environmental pressure. In the case of MFASR, such a pressure had been lacking since there had been no serious societal questioning of the Ministry's role as the country's lead agency in foreign affairs, which might explain the limited change of working procedures despite extensive application of advanced IT.

7.3.3 *Change management by senior leadership and experienced diplomats*

As the case of DFAIT demonstrates, implementation of IT had been greatly facilitated by a commitment to technology at the senior level in the organization. On the contrary, at the N-MFA and MFASR, the lack of interest in the potential of IT to improve the efficiency of diplomatic operations at the senior level had proven to be hampering development of organization-wide standards for IT-supported work. This suggests that hierarchy may also serve as a factor driving change.

Moreover, as the cases of introducing the systems for electronic management of documents at DFAIT and N-MFA indicated, new IT-supported procedures and routines are more likely to be accepted throughout a foreign ministry if the process is managed by experienced foreign service officers rather than solely by tech-support staff and/or hired external consultants. This is, firstly, related to their deep inside knowledge of the specifics of foreign service work and procedures. More importantly, though, experienced foreign service officers playing the role of IT-champions can establish support and raise expectations regarding a new set of IT-supported procedures by using their informal contacts throughout the foreign ministry. This strategy of *complementing a top-down push* of new IT-procedures and routines *with a users' pull* was applied by change managers at DFAIT. In addition to attempting to achieve the same effect through informal contacts throughout the organization, the team of change managers at N-MFA were also working with middle managers in developing new IT-supported practices and procedures; the main rationale of such a strategy being that in some years most of these middle managers will be at senior posts within the foreign ministry and IT-supported procedures will hence be firmly planted throughout the organization.

At MFASR, neither of these principles of change management had been applied to any significant degree. Implementation of IT had been led almost exclusively by the technical experts. The senior leadership and senior foreign service officers had been reluctant to invest time and energy into change management. The project of broad-based organizational transformation including a comprehensive review and modernization of IT-infrastructure entitled TREFA and launched in early 2007 was steered directly by the Head of Administration and seemed to enjoy also a high level of support from the foreign minister. The actual effects of this comprehensive reform would need to be investigated.

7.4 *Theoretical Observations*

7.4.1 *The diplomatic logic of appropriateness as a meta-driver of change*

As the findings from the three cases studied here indicate, to the extent that there was a sense of performance crisis at the foreign ministries (and this concerns mainly DFAIT and to an extent N-MFA), it was related to the notion of decreasing ability to live up to the expectations of being the lead agency in the country's foreign affairs. Similarly, information security procedures maintaining a high level of secrecy had been relaxed with the aim of enabling improved access to updated informational resources for foreign service staff and hence improve their ability to provide qualified foreign policy advice and decisions on behalf of their government. Finally, change management in foreign ministries had the same goal—to apply IT to improve the ability of the organization to promote state interests effectively. In general, the meta-drivers of change were the expectations related to the role of the foreign ministries as the *authorizers* of information pertaining to the foreign affairs of their state. In other words, changes were motivated by the increasing gap between the expectations related to the role of the diplomat and the decreasing ability to live up to that role due to rapidly changing conditions in the environment in which that role was being performed. As one of the reformists at DFAIT had argued recently, "We must find a way to retake the high ground" (Copeland 2005). This is one way of expressing what March and Olsen (2005:14) call the "internal aspiration level pressure for change caused by enduring gaps between institutional ideals and institutional practices." In other words, the evidence from the current study shows that *IT-supported change dynamics has been driven by the established logic of appropriateness informing diplomacy*. This indicates that institutional stability does not preclude change. On the contrary, institutional stability sometimes requires change. This is at odds with some of the accounts of institutions as bulwarks of inflexibility, inefficiency and resistance to change. As a powerful driving force of change, logic of appropriateness leads to path-dependent adaptation providing institutional stability in changing environmental conditions. The findings from the current study hence support the view of Eisenstadt (1964) that institutions allow for and even catalyze particular avenues of change.

7.4.2 *Institutional stability of diplomacy through diversity of organizational adaptation in foreign ministries*

The findings presented in this study suggest that there are variations in how foreign ministries in different countries react to the same stimulus—the information revolution leading to implementation and use of IT in governmental administrations. The archetype features of a traditional foreign ministry as the organizational carrier of diplomacy were shown to be renewed and to some extent transformed in the case of DFAIT, while the magnitude of organizational change due to IT-effects was lower in the case of N-MFA and there was almost no change in relation to the archetype in the case of MFASR. In fact, the latter case was showing a tendency of institutionalizing features that characterize the institutional archetype of foreign ministry as identified in chapter 2. IT-related organizational change in the three foreign ministries had been taking different paths leading to differences in working procedures, organizational structures and cultures in the three organizations. It seems that implementation and use of IT had led to a different modus of organizing diplomacy in Canada compared to Norway and Slovakia. Does this indicate a fragmentation of the institutional basis of diplomacy?

What was quite clear from the three cases is that every foreign ministry had been adapting to organizational requirements of their domestic national administrative systems. The role of DFAIT in relation to its national government had been significantly different than the role of N-MFA and MFASR in their governments. The federal structure of Canada, its gradual emergence as a sovereign nation and its inherent character as a contested state, led to a specific role of DFAIT as a support agency of governmental and provincial efforts abroad with a special emphasis on issues of trade rather than an archetypical high-politics-infused top-coordinator of a country's foreign affairs. Given the specifics of Canada's statehood, it is quite clear that having a foreign ministry of a traditional kind found in states of Continental Europe or in Japan was in fact never an option. Given this organizational background of DFAIT and its position within the Canadian government, it then seems natural that the organization was adopting IT-supported standards and procedures that enabled them to sustain their role as the main support agency providing the infrastructure for the operations of the Canadian government abroad. While IT-supported procedures of sharing information internally and externally, and use of the web for broad-based public debates of foreign policy might seem innovative from the perspective of the traditional archetype of a foreign ministry, they represent merely a continuation of existing political practice in Canada (e.g. tradition of foreign policy dialogues with Canadian citizens since the 1970s) in a new IT-supported form.

Similarly, the N-MFA's and MFASR's implementation and use of IT reflected their position within the Norwegian and Slovak governments as *the* lead agencies in charge of foreign affairs of their respective state. The specifics of the embeddedness of diplomacy within the respective national administrative context had been determining the avenues of IT-supported organizational adaptation at the respective foreign ministry. Locally appropriate adaptation enables each of the organizations to generate resources and legitimacy within their domestic national administrative system. If a foreign ministry would single-handedly start implementing structures, procedures and processes inconsistent with the standards adopted within their national administrative system, it would find it difficult to receive financial and other resources and would eventually be ineffective. Through these locally-specific adaptations, which lead to continuity of locally appropriate organizational arrangements for organizing diplomacy, stability of diplomacy as a global institutional order informing relations between states is sustained. Diversity in adaptations among foreign ministries is hence a necessary precondition for the maintenance of diplomacy as an institutional order. At a more general level one could hence propose that *an institution has best chances of remaining stable if it is embedded in various locally-appropriate organizational carriers*; i.e. institutional stability is ensured through continuation of diversity of the institutional organizational basis. On the contrary, forced isomorphic change leading to homogenization irrespective of local organizational contexts may lead to undermining of local equilibriums and eventually to the loss of legitimacy of the institution and to change of the institution and/or its increasing irrelevance.

For a diversity of adaptations of the organizational basis of an institution to be possible, a set of shared institutional principles ensuring maintenance of unity of the institutional order is necessary, i.e. an institutionalized universalism of principles accommodating difference. Coming back to the cases studied here, one can note that although each of the foreign ministries played a different role within their respective political system supported by a different set of organizational structures and procedures, none of the three organizations had abandoned the notion that the state is the main actor in the international system and information exchanged between states is to be authorized, may be held secret and the public may not be consulted on particular foreign policy issues in advance. Even DFAIT, in which IT-supported network structures and procedures of coordination and information sharing had become pervasive throughout the organization, did not abandon hierarchy as a principle assuring political accountability; secrecy as a norm assuring protection of sensitive information in networked exchanges; and indeed a degree of selectiveness in terms of what issues were put up for public debate and consultation, where some issues such as bilateral relations with countries representing the traditional diplomatic agenda were avoided.[4] Hence, the core organizing principles of diplomacy as an institution remain stable contributing to the maintenance of the role of the state as the primary unit of political organization in the world. A real challenge to diplomacy as an institution would come about if these principles were challenged and if the role of the state would hence be undermined. Although IT enables greater openness of foreign policy processes and greater inclusion of non-state actors, as was argued above, effects of IT are mediated by institutional settings of foreign ministries, and moreover, in and of itself IT does *not* carry a specific institutional logic of its own. However, coupled with political processes introducing alternative diplomatic logic of appropriateness—alternative notions of who is a diplomat and what represents appropriate diplomatic situations and actions may be introduced challenging the established institutional order resting upon states. The Ottawa-process leading to the ban on land-mines might be an example of such a process (see Price 1998, Cameron et al. 1998, Axworthy 2003, Holsti 2004).

---

[4] This point is well illustrated by the challenges in providing timely information to the domestic public as part of the 1983 *Access to Information Act* in Canada. See Chapter 4 for more details.

7.4.3 *The social structure of diplomacy as a buffer against environmental turbulence*

The choice to include MFASR in the current study as a case of a new foreign ministry founded amidst the set of developments often referred to as the information revolution was motivated by the expectation to find the organization permeated by new IT-supported principles for organizing diplomacy. As was shown in Chapter 6, this expectation was not warranted and the findings at MFASR instead indicated that the organization was in the process of adopting and institutionalizing norms, rules and working procedures that closely resemble the traditional model of a foreign ministry strictly following bureaucratic procedures and hierarchical lines of authority, developing a culture of secrecy, and leaning towards reluctance to get involved in broad public debates.

These findings are at odds with some of the organization theory, which suggests that organizations adopt those structures, procedures and technologies that are prevalent in their organizational environment and enable them to function most efficiently in such environment (Hannan and Freeman 1977, 1989, Hannan and Carol 1995). The point which the MFASR demonstrates so vividly, is that belonging to a well-established and entrenched institutional order (in this case the inter-state institutional order of diplomacy) provides organizations with legitimacy and buffers them from selection and other environmental pressures, which non-institutionalized organizations are normally faced with. As Meyer and Rowan ([1977]1991:53) argued, "independent of their productive efficiency, organizations which exist in highly elaborated institutional environments and succeed in becoming isomorphic with these environments gain the legitimacy and resources needed to survive." The study of foreign ministries prompts further theoretical thinking here.

Foreign ministries are embedded not only in the institutionalized system of the respective national government, but more profoundly also within the global diplomatic system. What is interesting about the inter-state diplomatic order compared with national administrative systems is that while the latter have been subject to ongoing changes in most states in recent centuries, the inter-state diplomatic order has changed fairly little over the last four hundred years (Mattingly 1955). The organizational form that foreign ministries have today can clearly be traced to the 17th Century French foreign ministry (Anderson 1993, Hamilton and Langhorne 1995). This may be related to the fact that the Westphalian state order as the *social structure*[5] of foreign ministries remains stable. Hence, for instance the N-MFA founded in 1905 or the MFASR founded in 1993 were essentially legitimized simply by becoming isomorphic with the prevalent models in the institutionalized environment of inter-state diplomacy, which was formed and stabilized in the process of state formation in 17th and 18th Century Europe.

In search of legitimacy, organizations are becoming isomorphic with current environments, but also with environments and constellations of legitimacy generating resources formed in different historical periods but still existing today—*surviving social structures*. This may lead to clashes of sets of norms, rules and procedures carried by the organizational fields of different ages. What may have been legitimate forms of organization and conduct three hundred years ago may not be as legitimate today. Each foreign ministry in a democratic state is hence exposed to isomorphic pressures from two organizational fields carrying different logics of appropriateness. On the one hand there is the organizational field of national governments with its norms of openness and citizen involvement in governance matters. On the other hand there is the inter-state organizational field of diplomacy with its norms of authorized information exchange, secrecy and low public involvement. Foreign ministries are hence in continuous need of accommodation of these two competing norm-sets and logics of appropriateness.

7.5 *Revolution Ante Portas?*

Even though the findings of the current study suggest that the effects of IT had not led to an IT-revolution in diplomacy, but a path-dependent adaptation leading to renewal of diplomacy, it is still an early assessment. As several authors argue (see for instance Heckscher and Donnelon 1994, Snellen and Van de Donk 1998, Bekkers 1998, Frissen 1997, 1999, Andersen 1999, Fountain 2001, Kamarck and Nye 2002, Baldersheim and Øgård 2003), the process of technological modernization in the governmental sector is at an early stage itself, and technology-supported shifts in governmental organizing, driven by the rise of government-wide portals for service provision undermining traditional bureaucratic

[5] Stinchcombe (1965:142) defines social structure as "groups, institutions, laws, population characteristics, and sets of social relations that form the environment of the organization (Stinchcombe 1965:142).

divisions between governmental ministries, might lead to fundamental change in the way governance within states is organized. It might be of interest to explore the impacts this emerging set of developments would have on the way diplomacy is organized within states.

As this study highlights, implementation and use of IT and the ensuing change at foreign ministries were motivated and limited by the logic of appropriateness informing diplomacy and foreign ministries as its carriers. Change was driven by the foreign ministries' decreasing capability to live up to their institutionalized role as guardians of the informational prerogative of the state in foreign affairs (see Figure 1 in Chapter 2). As long as this remains the case, foreign ministries will continue to play the role of the bastions of state interests. Involvement of non-state actors and the ensuing need for IT-supported network forms of organization will remain a rhetorical and symbolic claim without a transforming impact on the core organizing principles of diplomacy or its organizational basis.

Existence of states decreases the complexity of interactions among societies around the globe by providing a regularized system of information flows embodied in diplomacy. As in other areas of political organization, so in diplomacy too, the state has proven to be an efficient organizing unit and sovereignty an efficient organizing logic. However, as Krasner (1988:67) points out, given the omnipresent possibility of war between states and the potentially disastrous consequences of this systemic feature of the interstate system in the nuclear age, "it is no longer obvious that the state system is the optimal way to organize political life." Development of alternative principles of political organizing complementing sovereignty found in the works by academics (e.g. Schmitter 1996, Habermas 1998, Ansell and Weber 1999, Fossum 2002, Keohane 2002, Jacobsson and Sahlin-Andersson 2003, Krasner 2004), by practitioners and governments (e.g. Axworthy 2003, Cooper 2003), and indeed in actual day-to-day political processes such as those in the EU, are hence of interest as their adoption and diffusion might have transforming effects on the interstate system.

The availability of a different constitutive logic of action in the international system complementing or replacing sovereignty and introducing different notions of who are legitimate diplomatic actors in the global environment and what are legitimate diplomatic actions in particular situations, will inspire foreign ministries to harness the potential of IT in ways that would enable them to start playing the new role that the new organizing principles in the interstate system would entail. It is this shift in the constitutive logic of the international system that would lead to a transformation of diplomacy which would then feature a different configuration of legitimate actors and situations, and indeed a different organizational basis embedded in a different set of organizational arrangements. This would be a new logic of diplomatic appropriateness maintained by a new organizational field of diplomacy.

Following the IT-revolution, the technological infrastructure that would enable a revolution in organizing diplomacy is in place. What remains to be seen is whether the ideas introducing new notions of political organizing beyond state sovereignty and hence enabling such a revolution will take root.

REFERENCES

Allern, E. H. and Lorentzen, H. (2002): "New Forms of Civic Participation in Norwegian Government Administration. An Exploratory Study." Working Paper 2002:113, Oslo: NIBR.

Andersen, K. V. (1999): "Reengineering Public Sector Organizations Using Information Technology" in Heeks, R. (ed.): *Reinventing Government in the Information Age. International Practice in IT-enabled Public Sector Reform*. London: Routledge.

Anderson, M. S. (1993): *The Rise of Modern Diplomacy 1450–1919* London: Longman.

Ansell, C. K. and Weber, S. (1999): "Organizing International Politics: Sovereignty and Open Systems" International Political Science Review, 20 (1), pp. 73–93.

Avgerou, C. (2001): "The Significance of Context in Information Systems and Organizational Change" *Information Systems Journal*, Vol. 11, pp. 43–63.

Axworthy, L. (2003): *Navigating the New World. Canada's Global Future.* Toronto: Alfred A. Knopf.

Baldersheim, H. (2004): "In the Virtual Town Hall: New Information Regimes and a Third Face of Power? Nordic Cities on the Web" paper presented to the conference City Futures, University of Illinois at Chicago, July 8–10, 2004.

Baldersheim, H. and Øgård, M. (2003): *IKT på tvers av kommunegrenser* Oslo: Kommuneforlaget.

Balloch, H. and Angell, D. (1992): "Foreign Policy Planning and Coordination in Canada: The Policy Planning Staff at External Affairs" in *Canadian Public Administration*, 35 (4), pp. 449–463.

Barley, S. R. (1986): "Technology as an Occasion of Structuring: Evidence from Observations of CT Scanners and the Social Order of Radiology Departments" *Administrative Science Quarterly*, Vol. 31, pp. 78–108.

——. (1990): "Alignment of Technology and Structure through Roles and Networks" in *Administrative Science Quarterly*, Vol. 35, pp. 61–103.

Bátora, J. (2000): "Challenges and Opportunities: Slovak Diplomacy in the Information age." in *Slovak Foreign Policy Affairs*, 1 (1), pp. 86–101.

——. (2003): "The Institutional Background of Slovakia's Foreign Policy" in Kollár, M. and Mesežnikov, G. (eds.): *Slovakia 2003: Global Report on the State of Society* Bratislava: Institute of Public Affairs.

——. (2005): "Does the European Union Transform the Institution of Diplomacy?" *Journal of European Public Policy*, 12 (1), pp. 44–66.

——. (2006a): "Public Diplomacy between Home and Abroad: Norway and Canada" in *The Hague Journal of Diplomacy*, 1(1): 53–80.

——. (2006b): "Emerging Tenets of Responsive Foreign Policymaking?" in *International Journal*, 61: 929–942.

Bátora, J. and Neumann, I. B. (2002): "Cautious Surfers: The Norwegian Ministry of Foreign Affairs Negotiates the Wave of the Information Age" *Diplomacy and Statecraft*, 13 (3), pp. 23–56.

Baum, J. (1996): "Organizational Ecology" in Clegg, S. (ed.): *Handbook of Organizational Studies* London: SAGE Publications.

Bekkers, V. J. J. M. (1998): "Wiring Public Organizations and Changing Organizational Jurisdictions" in Snellen, I. Th. M. and Van de Donk, W. B. H. J. (1998): *Public Administration in an Information Age: A Handbook* Washington DC: IOS Press.

Bell, D. (1974): *The Coming of Post-Industrial Society: A Venture in Social Forecasting* London: Heinemann.

Berridge, G. R. ([1995] 2001): *Diplomacy—Theory and Practice.* 2nd ed. London: Palgrave.
Bijker, W. E., Hughes, T. P. and Pinch, T. (eds.) (1987): *The Social Construction of Technological Systems: New Directions in the Sociology and History of Technology* Cambridge, MA: MIT Press.
Bijker, W. E. and Law, J. (1992): *Shaping Technology/Building Society. Studies in Sociotechnical Change.* Cambridge, MA: MIT Press.
Blackwell, A. (2007): "Results-Based Diplomacy" in Rana, K. and Kurbalija, J. (eds.): *Foreign Ministries. Managing Diplomatic Networks and Optimizing Value.* Geneva: Diplo Books, pp. 47–54.
Blichner, L. C. and Sangolt, L. (1993): "Internasjonalisering av offentlig forvaltning" in Lægreid, P., Olsen, J. P. and Baldersheim, H. (eds.): *Organisering av offentlig sektor. Perspektiver, reformer, erfaringer, utfordringer.* Oslo: TANO, pp. 171–199.
Buchanan, D. and Boddy, D. (1992): *The Expertise of a Change Agent: Public Performance and Backstage Activity* New York: Prentice Hall.
Bucher-Johannessen, B. (1999): "Den norske modellen: Stat og samfunn hånd i hånd til fremme av Norge" *Internasjonal Politikk,* 57 (2), pp. 199–220.
Burchill, R. (1993): "Recent Structural Change in the Department" in Story, Donald C. (ed.): *The Canadian Foreign Service in Transition* Toronto: Canadian Scholars' Press
Castells, M. (1996): The Information age. Economy, Society and Culture, Vol. 1: *The Rise of the Network Society,* Oxford: Blackwell Publishers.
Cameron, A., Lawson, R. and Tomlin, B. (1998): *To Walk Without Fear. The Global Movement to Ban Landmines.* Toronto: Oxford University Press.
Cascone, A. (2000): "Comparing Diplomatic Services: Structures, Networks and Resources of the Ministries of Foreign Affairs of EU and G8 Member States." Occasional Paper, Malta: Diplo Foundation.
Christensen, T. (1996): *Adapting to Processes of Europeanisation. A Study of the Norwegian Ministry of Foreign Affairs.* Report No. 2/96 Oslo: ARENA.
Christensen, T. and Egeberg, M. (1989): "Norske departementer 1976–86: Noen trekk ved formell struktur, personell og beslutningsatferd" in Egeberg, M. (ed.): *Institusjonspolitikk og forvaltningsutvikling* Oslo: TANO.
Christensen, T. and Lægreid, P. (2001): *New Public Management: The Transformation of Ideas and Practice* Aldershot: Ashgate.
Clark, D. (2002): "Neoliberalism and Public Service Reform" in *Canadian Journal of Political Science,* Vol. 35, pp. 771–94.
Clemens, E. S. and Cook, J. M. (1999): "Politics and Institutionalism: Explaining Durability and Change" *Annual Review of Sociology,* Vol. 25, pp. 441–466.
Client/Server Case Study: Canada's Department of Foreign Affairs and International Trade (DFAIT) Rexdale, ON: Intel Corporation, 1994.
Cohen, W. and Levinthal, D. (1990): Absorptive Capacity: A New Perspective on Learning and Innovation" *Administrative Science Quarterly,* Vol. 35, pp. 128–152.
Cohen, R. and Westbrook, R. (2000): *Amarna Diplomacy: The Beginnings of International Relations* Baltimore: Johns Hopkins University Press.
Colding, K. (1978): *Organisasjonsspørsmål i Utenriksdepartementet,* Norsk Utenrikspolitisk Årbok 1978, Oslo: NUPI.
Coolsaet, R. (1998): "The Transformation of Diplomacy at the Threshold of the New Millennium" Discussion Papers in Diplomacy, No. 48 Leicester: Centre for the Study of Diplomacy.
Cooper, A. F. (1999): "Canada. Trying to Get it Right: the Foreign Ministry and Organizational Change" in B. Hocking (ed.): *Foreign Ministries. Change and Adaptation.* London: Macmillan Press, pp. 42–51.
——. (2004): *Tests of Global Governance: Canadian Diplomacy and United Nations World Conferences* New York: United Nations University Press.
Cooper, R. (2003): *The Breaking of Nations. Order and Chaos in the Twenty-First Century* New York: Atlantic Monthly Press.

Copeland, D. (2005): "New rabbits, old hats: International policy and Canada's foreign service in an era of reduced diplomatic resources" *International Journal*, 60 (3).

Deutsch, K. W. (1966): *The Nerves of Government: Models of Political Communication and Control*, New York: Free Press.

Deutsch, K. (1978): *The Analysis of International Relations* Englewood Cliffs, NJ: Prentice Hall.

DiMaggio, P. (1991): "Constructing an Organizational Field as a Professional Project" in Powell, W. W. and DiMaggio, P. (eds.): *The New Institutionalism in Organizational Analysis* Chicago: The University of Chicago Press, pp. 267–292.

DiMaggio, P. and Powell, W. W. (1991): "The Iron Cage Revisited: Institutional Isomorphism and Collective Rationality in Organizational Fields" in Powell, W. W. and DiMaggio, P. (eds.): *The New Institutionalism in Organizational Analysis* Chicago: The University of Chicago Press, pp. 63–82.

Dizard, W. (2001): *Digital Diplomacy. U.S. Foreign Policy in the Information Age.* Westport: Praeger.

e-diplomacy: the FCO e-business strategy. London: Foreign and Commonwealth Office, July 2001.

East, M. A. (1981): "The Organizational Effect of Interdependence on Foreign Policy Making. The Case of Norway." in Ch. W. Kegley and P. McGowan: *The Political Economy of Foreign Policy Behavior* Beverly Hills, CA: Sage Publications.

Eayrs, J. (1960): "The Origins of Canada's Department of External Affairs" in Keenleyside, Hugh et al. (eds.): *The Growth of Canadian Policies in External Affairs* London: Cambridge University Press.

——. (1971): *Diplomacy and its Discontents* Toronto: University of Toronto Press.

——. (1982): "Canada: The Department of External Affairs" in Steiner, Zara (ed.): *The Times Survey of Foreign Ministries of the World* London: Times Books.

Edge, D. ([1987] 1995): "The Social Shaping of Technology" in Heap, N. et al. (eds.): *Information Technology and Society* London: SAGE.

Egeberg, M. (1999): "The Impact of Bureaucratic Structure on Policy Making" *Public Administration* Vol. 77, pp. 155–170.

Egeberg, M. and Sætren, H.(1999): "Identities in Complex Organizations: A Study of Ministerial Bureaucrats" in Egeberg, M. and Lægreid, M. (eds.): *Organizing Political Institutions* Oslo: Universitetsforlaget, pp. 93–108

Egeberg, M., Christensen, T., Nikolaisen, R. and Stigen, I. M. (1989): "Departementene og andre sentraladministrative organer: En sammenligning av enkelte trekk ved formell struktur, personell og beslutningsadferd" in Egeberg, M. (ed.): *Institusjonspolitikk og forvaltningsutvikling* Oslo: TANO, pp. 54–74.

Eisenstadt, S. N. (1964): "Institutionalization and Change." *American Sociological Review* Vol. 29, pp. 235–247.

Eldon, S. (1994): *From Quill Pen to Satellite. Foreign Ministries in the Information Age.* London: The Royal Institute of International Affairs.

Equipped for the Future. Managing U.S. Foreign Affairs in the 21st Century (Report) Henry L. Stimson Center, Washington, DC., October 1998.

Fast, W. R. (1996): "Knowledge Strategies: Balancing Ends, Ways and Means in the Information Age", in Neilson, R. (ed.): *Sun Tzu Art of War in Information Warfare* Washington, DC: National Defense University (http://www.ndu.edu/inss/siws/ch1.html.)

Fossum. J. E. (2002): *The Transformation of the Nation-State: Why Compare the EU and Canada?* Working Paper No. 28/2002, Oslo: ARENA.

——. (2006): "Guest Editor's Introduction" in Gidsland: Is there a Mentor State? *International Journal* (special issue), 61(4):783–791.

Fountain, J. E. (2001): *Building the Virtual State: Information Technology and Institutional Change* Washington DC: Brookings Institution Press.

Friedland, R. and Alford, R. R. (1991): "Bringing Society Back In: Symbols, Practices and Institutional Contradictions" in Powell, W.W. and DiMaggio, P. (eds.): The New

Institutionalism in Organizational Analysis Chicago: The University of Chicago Press, pp. 232–263.
Frissen, P. (1997): "The Virtual State. Postmodernisation, Informatisation and Public Administration." In Loader, B. (ed.): *The Governance of Cyberspace. Politics, Technology and Global Restructuring.* London: Routledge, pp. 111–125.
——. (1998): "Public Administration in Cyberspace" in Snellen, I. T. and van de Donk, W. B. (eds.): *Public Administration in an Information Age. A Handbook.* Amsterdam: IOS Press, pp. 33–46.
Frissen, P. H. A. (1999): *Politics, Governance and Technology: A postmodern Narrative on the Virtual State* Cheltenham: Edward Elgar.
Garson, R. (2007): "Canada's Foreign Ministry: Online and Interactive" in Rana, K. and Kurbalija, J. (eds.): *Foreign Ministries. Managing Diplomatic Networks and Optimizing Value.* Geneva: Diplo Books, pp. 212–224.
Grant, R. (2005): "The Democratisation of Diplomacy: Negotiating with the Internet" *CDSP Working Paper*, The Hague: Clingendael Institute, September 2005.
Guehenno, J. M. (1995): *The End of the Nation-State* Minneapolis: University of Minnesota Press.
Gyárfášová, O. and Velšic, M. (2002): "Public Opinion" in Kollár, M. and Mesežnikov, G. (eds.): *Slovakia 2002. Global Report on the State of Society* Bratislava: Institute of Public Affairs.
Habermas, J. (1998): *The Inclusion of the Other* New York: Polity Press.
Hamilton, K. and Langhorne, R. (1995): *The Practice of Diplomacy—Its Evolution, Theory and Administration*, London: Routledge.
Hannan, M. T. and Carroll, G. R. (1995): An Introduction to Organizational Ecology" in Carroll, G. R. and Hannan, M. T. (eds.): *Organizations in Industry: Strategy, Structure and Selection* New York: Oxford University Press.
Hay, J. B. (2000): "Practising Democratic Foreign Policy: DFAIT's Consultations with Canadians" Ottawa: Canadian Centre for Foreign Policy Development.
Heckscher, C. and Donnellon, A. (1994): *The Post-Bureaucratic Organization: New Perspectives on Organizational Change* Thousand Oaks, CA: SAGE Publications.
Heeks, R. (ed.) (1999): *Reinventing Government in the Information Age. International Practice in IT-enabled Public Sector Reform.* London: Routledge.
Held, D. et al. (1999): *Global Transformations. Politics, Economics and Culture.* Cambridge: Polity Press.
Hill, C. (2003): *The Changing Politics of Foreign Policy*. London: Palgrave.
Hocking, B. (1999): "Foreign Ministries: Redefining the Gatekeeper Role" in Hocking, B. (ed.): *Foreign Ministries. Change and Adaptation.* London: Macmillan Press.
——. (2001): "Diplomacy: New Agendas and Changing Strategies" *Information Impacts Magazine*, July 2001.
——. (2007): "What is the Foreign Ministry?" in Rana, K. and Kurbalija, J. (eds.): *Foreign Ministries: Managing Diplomatic Networks and Optimizing Value.* Geneva: Diplo Books, pp. 3–19.
Holsti, K. J. (2004): *Taming the Sovereigns. Institutional Change in International Politics.* Cambridge: Cambridge University Press.
Høgetveit, E. (1981): *Hvor hemmelig? Offentlighetsprinsippet i Norge og USA, særlig med henblikk på militærpolitiske spørsmål.* Oslo: Pax Forlag AS.
Huntington, S. (1966): "Political Modernization: America vs. Europe" *World Politics*, 18 (3), pp. 378–414.
Hynek, N. (2008): "Conditions of Emergence and Their Effects: Political Rationalities, Governmental Programs and Technologies of Power in the Landmine Case" in *Journal of International Relations and Development* (forthcoming).
Ibsen, S. (1911 [1992]): *Videnskab og mystik og andre essays* Oslo: Grøndahl Dreyer.
Jones, R. A. (1983): *The British Diplomatic Service 1815–1914* Waterloo, ON: Wilfrid Laurier University Press.

Jones, D. T. (2005): "Doing Diplomacy Differently: The Canadian Foreign Service." in *Foreign Service Journal*, March 2005, pp. 43–50.

Kamarck, E. C. and Nye, J. S. (2002): *governance.com. Democracy in the Information Age.* Washington: Brookings Institution Press.

Keohane, R. (2002): "Ironies of Sovereignty: The European Union and the United States" in *Journal of Common Market Studies*, Vol. 40, No. 4, pp. 743–765.

Knudsen, O. F. (1997): "Beslutningsprosesser i norsk utenrikspolitikk" in Knutsen, T., Sørbø L., Gunnar M. and Gjerdåker, S. (eds.): *Norges utenrikspolitikk* Oslo: Cappelen Akademisk Forlag.

Krasner, S. (1988): "Sovereignty: An Institutional Perspective" in *Comparative Political Studies*, 21 (1), pp. 66–94.

——. (1999): "Logics of Consequences and Appropriateness in the International System" in Egeberg, M. and Lægreid, P. (eds.): *Organizing Political Institutions* Oslo: Scandinavian University Press, pp. 181–213.

——. (2004): "Sharing Sovereignty. New Institutions for Collapsed and Failing States" *International Security*, 29 (2), pp. 85–120.

Kristiansen, K. (2004): "Kontakte und Reisen im 2. Jahrtausend v. Chr." In *Rad und Wagen* Mainz am Rhehn, pp. 443–454.

Kurbalija, J. (2007): "E-Diplomacy: The Challenge for Ministries of Foreign Affairs" in Rana, K. and Kurbalija, J. (eds.): *Foreign Ministries: Managing Diplomatic Networks and Optimizing Value.* Geneva: Diplo Books, pp. 305–338.

Langhorne, R. and Wallace, W. (1999): "Diplomacy towards the 21st Century" in Hocking, B. (ed.): *Foreign Ministries: Change and Adaptation.* London: Macmillan Press, pp. 16–22.

Lauren, P. G. (1976): *Diplomats and Bureaucrats. The First Institutional Responses to Twentieth-Century Diplomacy in France and Germany* Stanford: Hoover Institution Press.

Loader, B. (ed.) (1999): *The Governance of Cyberspace. Politics, Technology and Global Restructuring.* London: Routledge.

Lukáč, P. (2004): "Historická a politická identita Slovenska na prahu jeho integrácie do EÚ" in Lukáč, P.: *Dejiny a zahraničná politika v strednej Európe* Bratislava: Kalligram, pp. 311–325.

Mackenzie, H. (1998): *Recruiting Tomorrow's Ambassadors: Examination and Selection for the Foreign Service of Canada, 1925–1997* Ottawa: DFAIT (www.dfait-maeci.gc.ca/hist/docs/converted/recruitment-e.htm).

March, J. G. and Olsen, J. P. (1989): *Rediscovering Institutions. The Organizational Basis of Politics.* New York: The Free Press.

——. (1995): *Democratic Governance*, New York: The Free Press.

——. (2005): "Elaborating 'New Institutionalism'" Working Paper 11/05, Oslo: ARENA; also in Rhodes, R. A. W., Binder, S. and Rockman, B. (eds.): *The Oxford Handbook of Political Institutions.* Oxford: Oxford University Press.

——. (2006): "The Logic of Appropriateness" in Moran, M., Rein, M. and Goodin, R. E. (eds.): *Oxford Handbook of Public Policy* Oxford: Oxford University Press, pp. 689–708.

Mattingly, G. (1955): *Renaissance Diplomacy* Oxford: Alden Press.

Melissen, J. (1999): *Innovation in Diplomatic Practice* Basingstoke: Macmillan.

——. (2005): "Wielding Soft Power: The New Public Diplomacy" *Clingenadael Diplomacy Paper*, No. 2, The Hague: Netherlands Institute of International Relations, May 2005.

Metzl, J. (2001): "Network Diplomacy" in *Georgetown Journal of International Affairs*, Winter/Spring 2001 (www.ceip.org/files/publications/network_diplomacy.asp).

Meyer, J. W. and Rowan, B. ([1977] 1991): "Institutional Organizations: Structure as Myth and Ceremony" in Powell, W. and DiMaggio, P. J. (eds.): *The New Institutionalism in Organizational Analysis* Chicago: The University of Chicago Press, pp. 63–82.

Michaud, N. (2001): "Federalism and Foreign Policy: Comparative Answers to Globalisation" paper presented to the conference "The Administration of Foreign Affairs: A Renewed Challenge?", Hull, Quebec, November 1–3, 2001.

Mojžita, M. (2004): *Kňažko/Demeš/Kňažko. Formovanie slovenskej diplomacie v rokoch 1990 až 1993* Bratislava: Ústav politických vied SAV.

Moses, J. W. and Knutsen, T. (2001): "Inside Out. Globalization and the Reorganization of Foreign Affairs Ministries" in *Cooperation and Conflict*, Vol. 36, No. 4, pp. 355–380.

Negroponte, N. (1995): *Being Digital* London: Hodder & Stoughton.

Neumann, I. B. (1998): "Departemental identitet: Det norske utenriksdepartement" [Ministerial Identity: The Norwegian Foreign Ministry] in *Internasjonal Politikk* 56 (1), pp. 75–103.

——. (1999): "Norsk sørpolitikk: den disaggregerte stats diplomati" *Internasjonal Politikk*, Vol. 57, No. 2, pp. 181–197.

——. (2001a): "'Grab a Phaser, Ambassador': Diplomacy in Star Trek" in *Millennium*, Vol. 30, No. 3, pp. 603–624.

——. (2001b): *Mening, materialitet, makt* Bergen: Fagbokforlaget

——. (2001c): *Diplomati som historisk og sosial praksis* MA thesis in social anthropology, University of Oslo.

——. (2005): "Sett UD modernisering på timeplanen" *Bistandsaktuellt*, 03.03.2005.

Neumann, I. B. and Leira, H. (2005): *Aktiv og avventende: utenrikstjenestens liv 1905–2005* Oslo: PAX.

Nicolson, H. ([1939]1988): *Diplomacy* Washington, DC: Institute for the Study of Diplomacy, Georgetown University.

——. ([1953]1977): *The Evolution of the Diplomatic Method* Westport, CN: Greenwood Press Publishers.

Nič, M. (2007): "Svet zaseknutý v čase", interview in *OS—Občianská spoločnost'*, 5/2007, pp. 31–42.

Norman, E. W. (1982): "Norway" in Steiner, Zara (ed.): *The Times Survey of Foreign Ministries of the World* London: Times Books, pp. 391–405.

Nossal, K. R. (1993): "Contending Explanations for the Amalgamation of External Affairs" in Story, Donald C. (ed.): *The Canadian Foreign Service in Transition* Toronto: Canadian Scholars' Press.

——. (1997): *The Politics of Canadian Foreign Policy* 3rd ed., Scarborough, Ont.: Prentice Hall.

——. (2001): "The Administration of Provincial International Activities: Provincial Missions Abroad" paper presented to the conference "The Administration of Foreign Affairs: A Renewed Challenge?", Hull, Quebec, November 1–3, 2001.

Nye, J. S. (2002): *The Pradox of American Power* New York: Oxford University Press.

Olsen, J. P. (1992): "Analyzing Institutional Dynamics" *Staatswissenschaften und Staatspraxis*, Vol. 3, No. 2, pp. 247–271.

——. (2005): *The Institutional Dynamics of the (European) University* Working Paper 15/2005 Oslo: ARENA.

——. (2008): "Understanding Institutions and Logics of Appropriateness: Introductory Essay" in March, J. G.: *Understanding Organizations*. Stanford, CA: Stanford University Press (forthcoming).

Olsen, J. P. and Sverdrup, B. O. (1997): *Samarbeid og integrasjon—i Norden og i Europa* Working paper 97/20, Oslo: ARENA.

Omang, R. (1954): *Utenrikstjenesten* Oslo: Gyldendal Norsk Forlag.

Orlikowski, W. J. (1992): "The Duality of Technology: Rethinking the Concept of Technology in Organizations" *Organization Science*, Vol. 3, pp. 398–427.

——. (1996): "Improvising Organizational Transformation over Time: A Situated Change Perspective" *Information Systems Research*, Vol. 7, pp. 63–92.

Orren, K. and Skowronek, S. (2004): *The Search for American Political Development* Cambridge: Cambridge University Press.
Parkhurst-Ferguson, P. (1998): "A Cultural Field in the Making: Gastronomy in 19th Century France" *American Journal of Sociology*, 104 (3), pp. 597–641.
Peško, M. (2007): "Slovenská diplomacia sa reformuje" in *Zahraničná politika*, 3/2007.
Peters, B. G. (1999): Institutional Theory in Political Science: The 'New Institutionalism' London: Continuum.
Potter, E. H. (2001): "Information Technology and Canada's Public Diplomacy" paper presented to the conference "The Administration of Foreign Affairs: A Renewed Challenge?", Hull, Quebec, November 1–3, 2001.
——. (2002): *Cyberdiplomacy: Foreign Policy in the 21st Century.* Montreal and Kingston: McGill-Queen's University Press.
Potter, E. H. and Bernier, L. (2001): *Business Planning in Canadian Public Administration.* Toronto: Institute of Public Administration of Canada.
Preston, P. (2001): *Reshaping Communications. Technology, Information and Social Change.* London: SAGE.
Price, R. (1998): "Reversing the Gun Sights: Transnational Civil Society Targets Land Mines" in *International Organization*, Vol. 52, No. 3, pp. 613–644.
Queller, D. E. (1967): *The Office of the Ambassador in the Middle Ages* New Jersey: Princeton University Press.
Rana, K. (2007): "MFA Reform: Global Trends" in Rana, K. and Kurbalija, J. (eds.): *Foreign Ministries: Managing Diplomatic Networks and Optimizing Value.* Geneva: Diplo Books, pp. 20–46.
Reinventing Diplomacy in the Information Age (Report) Washington, DC.: Center for Strategic & International Studies, October 9, 1998 (released November 1, 1998).
Reus-Smit, C. (1999): *The Moral Purpose of the State: Culture, Social Identity and Institutional Rationality in International Relations.* Princeton, NJ: Princeton University Press.
Rice, C. (2006): "Transformational Diplomacy" speech delivered at Georgetown University, Washington, DC, January 18, 2006 (www.state.gov/secretary/rm/2006/59306.htm).
Ringmar, E. (1996): *Identity, Interest and Action. A Cultural Explanation of Sweden's Intervention in the Thirty Years War* Cambridge: Cambridge University Press.
Robey, D. and Bodreau, M. C. (1999): "Accounting for the Contradictory Organizational Consequences of Information Technology: Theoretical Directions and Methodological Implications" *Information Systems Research*, Vol. 10, No. 2, pp. 167–184.
Rosenau, J. (1998): "States, Sovereignty, and Diplomacy in the Information Age." paper presented to the Annual Meeting of the International Studies Association, Washington, DC., February 18, 1998; available online at http://www.usip.org/virtualdiplomacy/publications/reports/jrosenauISA99.html.
Rosenberg, H. (1958): *Bureaucracy, Aristocracy and Autocracy. The Prussian Experience 1660–1815* Boston: Beacon Press.
Ruggie, J. G. (1993): "Territoriality and Beyond: Problematizing Modernity in International Relations" *International Organization*, 40: 139–174.
Savoie, D. J. (1990): *The Politics of Public Spending in Canada* Toronto: University of Toronto Press.
——. (2004): "Searching for Accountability in a Government without Borders" in *Canadian Public Administration*, Vol. 47, pp. 1–26.
Schmitter, P. (1996): *Is it Really Possible to Democratize the Euro-Polity?* Working Paper No. 10/96 Oslo: ARENA.
Schmitz, C. A. (2001): "A Vision of an American Embassy in 2015: A Disquisition in Three Acts and Six and a Half Premises" *Information Impacts Magazine*, July 2001 (available online at http://www.usip.org/virtualdiplomacy/publications/reports/16.html#vision).
Scott, W. R. (2001): *Institutions and Organizations.* 2nd ed. London: Sage Publications.

——. (2004): "Reflections on a Half-Century of Organizational Sociology" *Annual Review of Sociology*, Vol. 30, pp. 1–21.
Sending, O. J. and Neumann, I. B. (2006): "Governance to Governmentality: Analyzing NGOs, States, and Power" *International Studies Quarterly*, Vol. 50, pp. 651–672.
Sharp, P. (1999): "Diplomacy: Representation and the Study of International Relations" *International Studies Review*, 1 (1), pp. 33–57.
Simon, H. A. (1998): "Information 101: It's Not What You Know, It's How You Know It" *The Journal of Quality and Participation*, July-August 1998, pp. 30–33.
Skilling, H. G. (1945): *Canadian Representation Abroad: From Agency to Embassy* Toronto: The Reyerson Press.
Slaata, T. (ed.) (2002): *Digital Makt. Informasjons- og kommunikasjonsteknologiens betydning og muligheter.* Oslo: Gyldendal.
Smith, G. S. (1997): "The Challenge of Virtual Diplomacy" paper presented to the Virtual Diplomacy Conference, April 1–2, 1997, Washington, DC. (http://www.usip.org/virtualdiplomacy/publications/papers/gsmith.html).
Smith, G. S. and Sutherland, A. (2002): "The New Diplomacy: Real-Time Implications and Applications" in Potter, E. H. (ed.): Cyber-Diplomacy. Managing Foreign Policy in the Twenty-First Century. Montreal: McGill-Queen's University Press.
Snellen, I. Th. M. and Van de Donk, W. B. H. J. (1998): *Public Administration in an Information Age: A Handbook* Washington DC: IOS Press.
Solheim, N. and Solem, K. A. (1997): "Rapport fra den digitale steinalder" [Report from the Digital Stone Age] in *UD-posten*, 2/97, internal bulletin of the Norwegian foreign ministry.
Spruyt, H. (1994): "Institutional Selection in International Relations: State Anarchy as Order" *International Organization*, 48(4): 527–557.
Stairs, D. (1973): "Canadian External Relations: Trends in Policymaking or The Building of the Rabbit Warren" paper prepared for a seminar at the Centre for Foreign Policy Studies, January 1973.
——. (2001): "The Administrative Politics of Canadian Foreign Policy: Past Realities and Current Trends" paper presented to the conference "The Administration of Foreign Affairs: A Renewed Challenge?", Hull, Quebec, November 1–3, 2001.
Steiner, Z. (1982): *The Times Survey of Foreign Ministries of the World* London: Times Books.
Stinchcombe, A. (1965): "Social Structure and Organizations." in March, J. G. (ed.): *Handbook of Organizations* Chicago: Rand McNally.
Strážay, T. (2007): "Western Balkans and the Visegrad Group: Toward a New Partnership?" in Bútora, M. et al. (eds.): *Democracy and Populism in Central Europe: The Visegrad Elections and Their Aftermath.* Bratislava: IVO, pp. 231–241.
Sundelius, B. (1984): "Interdependence, Internationalization and Foreign Policy Decentralization in Sweden." *Cooperation and Conflict*, Vol. 19, pp. 93–120.
The Electronic Foreign Service. Proposal for a strategy for information and communications technology and information management at the Norwegian Ministry of Foreign Affairs 2000–2003. Oslo: N-MFA, June 25, 1999 (available online at http://odin.dep.no/ud/english/doc/reports/032001–220005/dok-bn.html).
Thelen, K. (1999): "Historical Institutionalism in Comparative Politics" *Annual Review of Political Science*, Vol. 2, pp. 369–404.
Tilly, C. (ed.) (1975): *The Formation of National States in Western Europe* Princeton, N.J.: Princeton University Press.
Toffler, A. (1980): *The Third Wave* London: Collins.
Van Creveld, M. (1999): *Rise and Decline of the State* New York: Cambridge University Press.
Velšic, M. (2005): *Digitálna gramotnosť na Slovensku* Bratislava: Institute of Public Affairs.
Watson, A. (1982): *Diplomacy. The Dialogue Between States.* London: Eyre Methuen.

Weber, M. (1978): *Economy and Society: An Outline of Interpretive Sociology* Berkeley: University of California Press.
Wight, M. (1978): *Power Politics* Leicester: Leicester University Press.
Wriston, W. (1992): *The Twilight of Sovereignty* New York: Charles Scribner's Sons.
Yin, R. (1981): "The Case Study as a Serious Research Strategy" in *Knowledge: Creation, Diffusion, Utilization*, Vol. 3, No. 1, pp. 97–114.
——. (1994): *Case Study Research. Design and Methods* 2nd Edition, London: Sage.
Zuurmond, A. (1996): "From Bureaucracy to Infocracy: Towards Management through Information Infrastructure" in Taylor, J. A., Snellen, I. T. and Zuurmond, A. (eds.): *Beyond BPR in Public Administration. Institutional Transformation in an Information Age.* Amsterdam: IOS Press, pp. 205–224.
——. (1998): "From Bureaucracy to Infocracy: Are Democratic Institutions Lagging Behind?" in Snellen, I. T. and van de Donk, W. B. (eds.): *Public Administration in an Information Age. A Handbook.* Amsterdam: IOS Press, pp. 259–272.

APPENDICES

APPENDIX 1

INTERVIEWS AT THE CANADIAN DEPARTMENT OF FOREIGN AFFAIRS AND INTERNATIONAL TRADE

Date	*Name*	*Position*
November 2, 2001	Daryl Copeland	Director, South-East Asia Division
November 3, 2001	Evan H. Potter	Special Advisor (Communications), Policy Planning Division
November 6, 2001	Paul Chapin	Chair, Task Force on Modernizing the Foreign Service; Senior Coordinator, Universal Classification Standard Conversion Project
November 6, 2001	Gilles Gingras	Senior Advisor, Education Marketing Unit (formerly Counsellor and Consul at the Canadian Embassy in Oslo)
November 7, 2001	Gregory D. Graham	Director, Program Analysis and Budgeting Division
November 7, 2001	Lise Rivet	Trade Officer, Team Canada Inc, Trade Liaison Division
November 8, 2001	Joe R. Charlton	Director, Information Management and Technology Planning and Direction
November 8, 2001	Louis M. Gaëtan	Director, Client Services Division
November 8, 2001	Marie-Lucie Morin	Director General, International Business Development Policy and Planning
November 8, 2001	Karen MacArthur	Director, Government On-Line Project Office
November 9, 2001	Pierre Sabourin	Director, Export Development Division
November 5, 2003	Evan Potter and	Senior Advisor, Policy Planning
	Peter Glover and	Consultant, Web-Services
	Mark McLaughlin	eCommunications Strategist, Policy Planning
November 6, 2003	Daniel Marchand	Deputy Chief Information Officer

Appendix 1 (*cont.*)

Date	*Name*	*Position*
November 6, 2003	Joe R. Charlton	Director, Application Management Services
November 6, 2003	Jacques Desjardins	Deputy Director and Project Manager, 'InfoBank' Project
November 7, 2003	Pierre Sabourin	See above
November 12, 2003	Tricia McDonald	Chief, Public Diplomacy Program
November 13, 2003	Aubrey Morantz and	Director, Corporate Security Division
	Robert Steward	Deputy Director, IT-Security, Corporate Security Division
November 13, 2003	Daryl Copeland	Senior Policy Advisor, Communications Services Division
November 14, 2003	Robert Desjardins	Director, Japan Division
November 14, 2003	Barry Nesbitt	Strategist Government On-Line, outreach Programs and E-Communications Division
November 14, 2003	Josette Couture	Deputy Director, Outreach Programs and E-Communications Division

APPENDIX 2

INTERVIEWS AT THE NORWEGIAN MINISTRY OF FOREIGN AFFAIRS

Date	*Name*	*Position*
March 16, 1999	Tore Nedrebø	Head of Division of the Policy Planning and Evaluation Office
March 16, 1999	Erik A. Underland	Senior Executive Officer of the Section for Organizational Development and Information Management
March 17, 1999	Ove F. Mo	Director, ICT Section
March 17, 1999	Zaida Hessen	Senior Executive Officer/Project Manager, ICT Section
March 18, 1999	Carsten J. Helgeby	Director General of the Section for Information and Press
March 18, 1999	Narve Solheim	Senior Executive Officer of the Section for the Promotion of Norwegian Culture Abroad
October 22, 1999	Tore Nedrebø	See above
May 6, 2003	Hans O. Seljegard	Project-Coordinator, 'BEST' Project
May 6, 2003	Kjell T. Pettersen	Assistant Director General, Department for Administrative Affairs
May 7, 2003	Leidulv Namtvedt and Ove F. Mo	Director General, Department for Administrative Affairs See above
May 7, 2003	Cecilie Willoch	Senior Advisor, Section for Information and Press Relations
May 8, 2003	Cecilie Landsverk	Director General, Multilateral department
May 8, 2003	Espen Rikter-Svendsen	Assistant Director, Department of Press, Culture and Information
May 9, 2003	Terje Hauge	Director, Department for Administrative Affairs
May 9, 2003	Anne G. Webb	Senior Executive Officer, Archive division
May 26, 2003	Ove F. Mo	See above

Appendix 2 (*cont.*)

Date	*Name*	*Position*
May 27, 2003	Zaida Hessen	Senior Executive Officer, IT-training, 'BEST' Project
May 28, 2003	Tore Nedrebø	Director, Section for European Policy
November 12, 2003	Jannicke Jæger	Minister Counsellor, Royal Norwegian Embassy in Ottawa
November 24, 2004	Ole A. Lindeman and	Project Coordinator, 'BEST' Project
	Cecilie Willoch	Senior Advisor, Secretariat of the Minister of International Development

APPENDIX 3

INTERVIEWS AT THE MINISTRY OF FOREIGN AFFAIRS OF THE SLOVAK REPUBLIC

Date	*Name*	*Position*
October 20, 1998	Jaroslav Uhrín	Director, Department of Information Systems
October 30, 1998	František Havran	Director, Department for Special Tasks (information security)
October 30, 1998	Jozef Zelizka	Director, Department for Special Communication
November 26, 1998	Zuzana Benková	Head, Division for Training and Education, Personnel Department
November 26, 1998	Ivana Repková	Executive Officer, Personnel Department
December 1, 1998	Miroslav Wlachovský	Director, Department of Analyses and Planning
December 10, 1998	Jozef Čemes	Head of Division, Personnel Department
January 4, 1999	Jozef Zelizka	See above
August 9, 1999	Miroslav Wlachovský	See above
August 20, 1999	Jozef Zelizka	Director, Department of Information Technology and Special Communication
April 2, 2003	Peter Holásek	Director, Department of Analyses and Planning
April 2, 2003	Juraj Pavlovič and Martin Urbančok	Deputy Director, Legislature and Organization Department; Senior Advisor, Legislature and Organization Department
April 2, 2003	L'ubomír Golian	Director, Department of Cultural Relations and Expatriates
April 3, 2003	Miroslav Mojžita	Head, Civil Service Authority (highest civil servant of the Ministry)

Appendix 3 (*cont.*)

Date	*Name*	*Position*
April 3, 2003	Michal Slivovič	Advisor to the Head of the Civil Service Authority
April 3, 2003	Jozef Vrábel	Director, Department of Information Technology and Special Communication
April 3, 2003	L'ubomír Kopaj	Director General, Division for Administration, Logistics and Information Technology
April 4, 2003	Ivan Horský	Director, Press and Communication Department
April 4, 2003	Janka Stadtruckerová	Executive Officer, Press and Communication Department
April 10, 2003	Radovan Kollár and	Deputy Director, Department of the Archives, Documentation and Registry Administration
	Jozef Sadloň	Senior Advisor, same department
April 10, 2003	Karol Dendis	Executive Officer, Press and Communication Department
April 10, 2003	Ivan Tichý	Director, Department of the Archives, Documentation and Registry Administration
April 14, 2003	Milan Cigáň	Senior Advisor, Department of Analyses and Planning
April 15, 2003	Anton Jurkovič	Head, Human Resources Development Division, Personnel Department
April 17, 2003	Anna Gogová	Senior Secretary, Department of Analyses and Planning
April 17, 2003	Marta Dubayová	Senior Executive Officer, Division for European Affairs
April 17, 2003	Milan Cigáň	See above
April 22, 2003	Peter Burian	Senior Advisor, Department of Analyses and Planning (former ambassador to NATO)
November 11, 2003	Štefan Rozkopál	Chargé d'affaires a.i., Slovak embassy in Ottawa
May 5, 2004	Norbert Brada	Advisor-political affairs, Permanent Mission of the Slovak Republic to the UN, New York
August 3, 2004	Jozef Vrábel	See above

APPENDIX 4

DEPARTMENT OF FOREIGN AFFAIRS AND INTERNATIONAL TRADE—ORGANIZATION CHART (SOURCE: DFAIT, 2002)

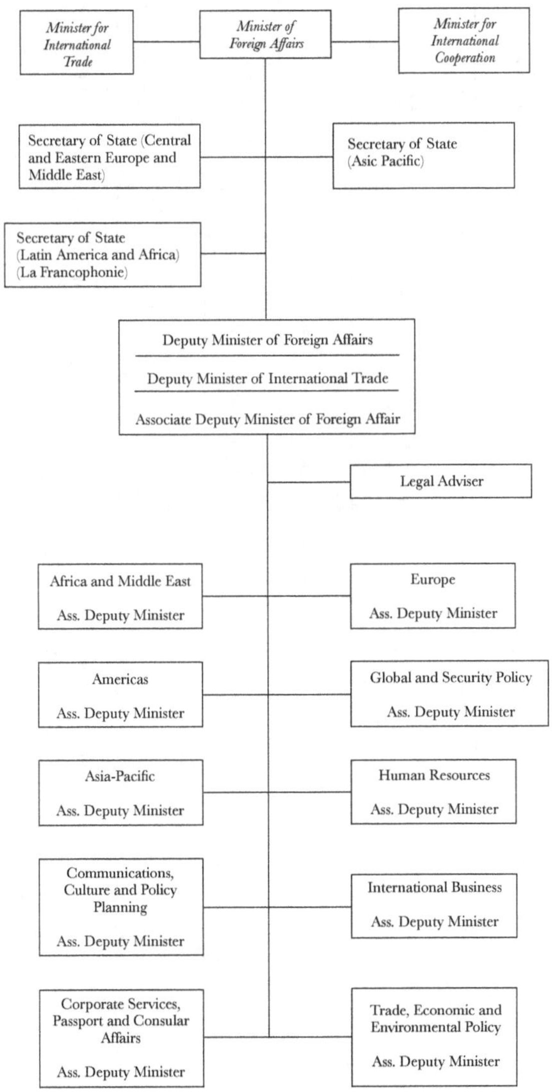

APPENDIX 5

THE ROYAL NORWEGIAN MINISTRY OF FOREIGN AFFAIRS—ORGANIZATION CHART (SOURCE: NEUMANN 1999:164)

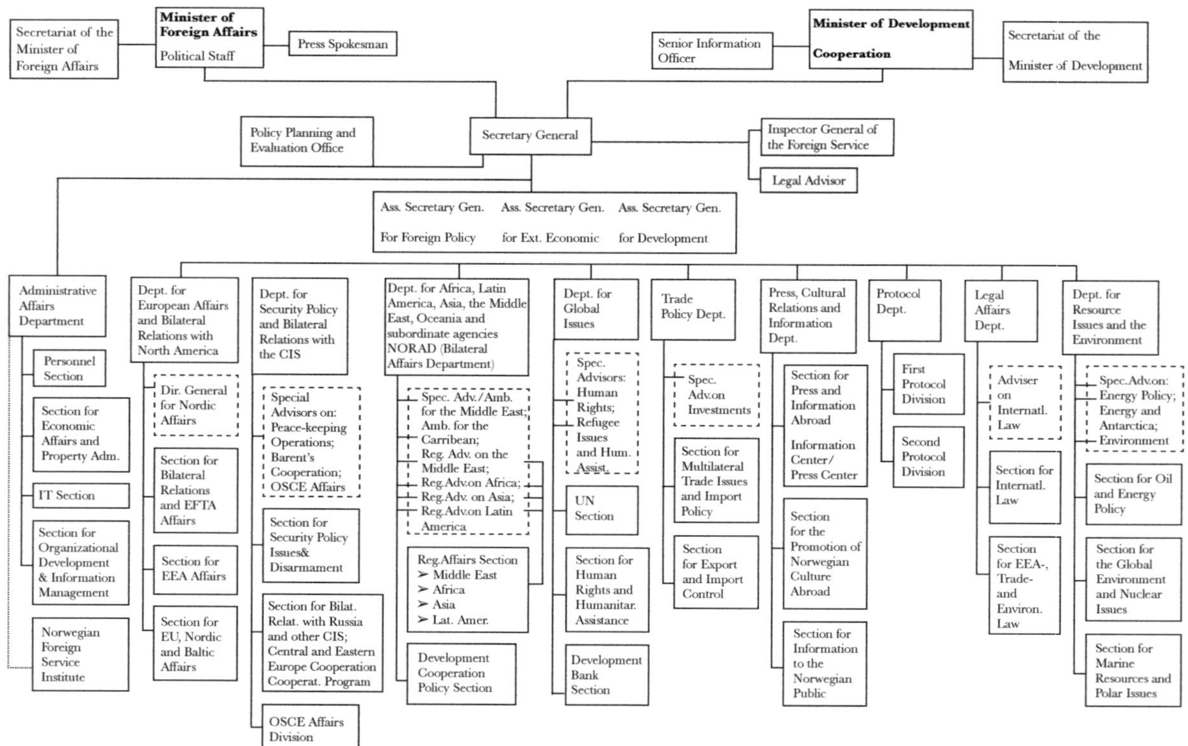

APPENDIX 6

THE MINISTRY OF FOREIGN AFFAIRS OF THE SLOVAK REPUBLIC—ORGANIZATION CHART (SOURCE: MFASR, 2001)

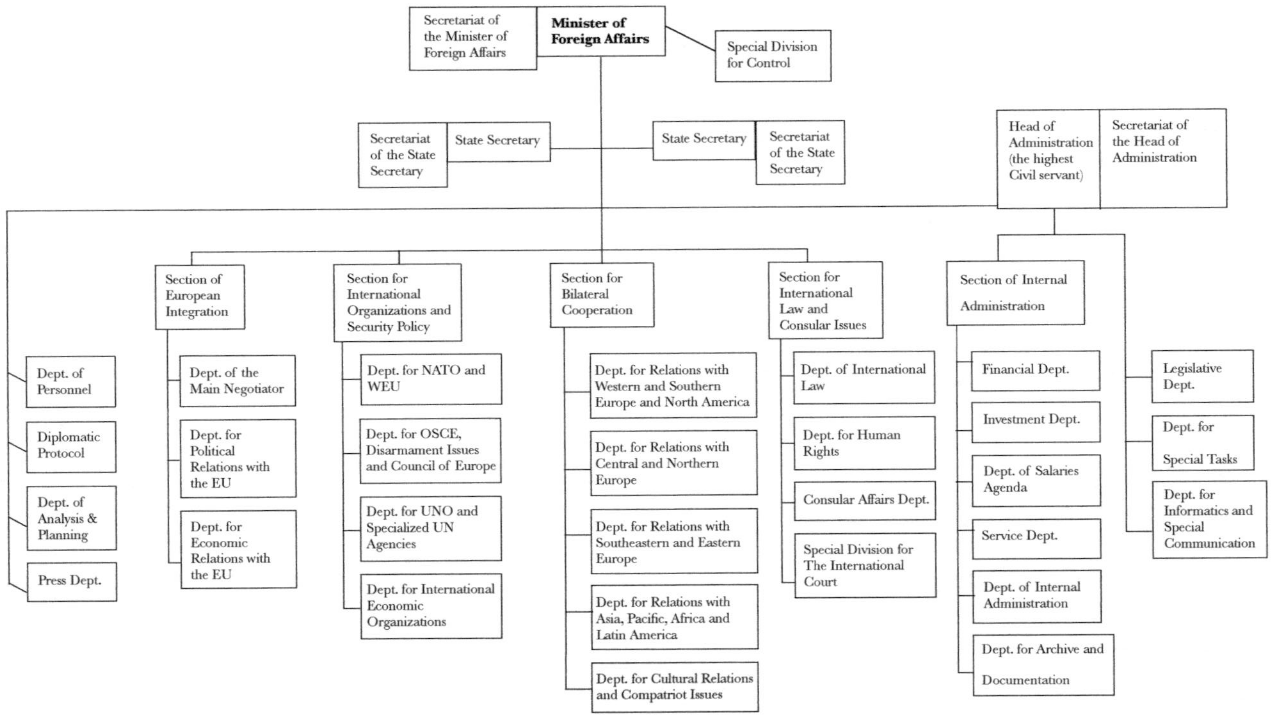

INDEX

www.ingramcontent.com/pod-product-compliance
Lightning Source LLC
LaVergne TN
LVHW011123300926
849160LV00008B/487